IMMUNE CROSSOVER IV

Network's Faces

The Lost Environment

Enrique Rewald

IMMUNE CROSSOVER IV

Network's Faces

The Lost Environment

Enrique Rewald

Co-Authors (Revised 2009)
Mercedes Francischetti
Pablo Alejandro Sánchez

Visit us online at www.authorsonline.co.uk

An Authors OnLine Book

ISBN 978 07552 0479 3

Authors OnLine Ltd
19 The Cinques
Gamlingay, Sandy
Bedfordshire SG19 3NU

This book is also available in e-book format, details of which are available at
www.authorsonline.co.uk

Dedicated to

Herbert Begemann [1919 - 1994]

Humane visionary and outsider

Everything in the universe interconnects,
Immunity being the exception.

Thoughts about collective interactive immunity:
an interdisciplinary speculative perspective.

ACKNOWLEDGEMENTS

In first place we wish to highlight the creative and most intelligent cooperation by Pablo Ignacio Martin, as hard working trainee he coauthored the 2007 edition, 'A propos time and autoimmunity', Clin Rev Allergy Immunol 2008:380-4 (Review); 'The other side of the coin: could microparticles serve as an interindividual immune link?', Scand J Immunol 2008; 67:103; 'Autoimmunity and time entanglement', Ann N Y Acad Sci. 2007;37-9 and, importantly, 'Microparticles and the Hygiene Hypothesis', an extended review in the forthcoming issue from the Ann NYAS. We wish that the winning streak will accompany him during his engineer career. The occasion also is proper to mention the professionalism and excellent 'chemistry' displayed by Phil Clinker, our proofreader and sub-editor. Further, we must state the sponsorship by Dr. Lazar y Cía S.A.Q. e I. My gratitude extends to my friend Peter Terness (Heidelberg University), who complemented his 'Foreword' with some most valuable advice. I wish to express again my appreciation to Professor Alberto de la Torre from Universidad Nacional de Mar del Plata. And I'll never forget the help given to me by the late physicist José Kleiner [1910–1997] – a former pupil of Albert Einstein and Max Planck – and my late secretary Ana María Harvey. I appreciate the immunohematological insight and expertise of Dr. Carlos Alberto González and the revision of previous editions by Dr. José Cabero. Last but not least, the goodwill of Mr Richard and Mrs Marjorie Fitt deserves my appreciation. Finally, I wish to express again my very special feelings of gratitude to Margarita, my wife.

CONTENTS

FOREWORD

A long time ago, when the whole earth was of one language, an ambitious humanity congregated in Shinar (Babylon) and set about building a great tower, a colossal staged temple that would reach heaven, so the Old Testament teaches us. This plan was in direct disobedience to Jehovah's command to fill the earth, and God was displeased with the unholy enterprise. He declared: "Come, let us go down, and there confound their language, that they may not understand one another's speech" (Gen. 11:7). The sacred decree was implemented and human languages were born.

Evolution of mankind has allowed unexpected achievements – who would have expected that only 50 years ago bacteria could be induced to produce human proteins, or that a sheep could be cloned? – but the price we have to pay is an increasing specialization, leading to experts for the one or other disease, organ, cell, or even molecule. It seems that God's curse of confusing human beings by giving them various languages has become reality. Just put a statistician together with a biologist and an historian and make them speak the same language! What humanity will need during the next decades are experts who, beyond their special fields, have not lost their perspective and try to integrate what specialization has divided.

Enrique Rewald's book is an attempt to look beyond microbes, the immune system and man; it tries to analyse the interplay among all these actors, each of which constitutes the special topic of experts. Apart from that, the author proposes original, often provocative or even contestable hypotheses on how these systems interact, with a special focus on the relationship between microbes and humans. When appraising such a monograph, one should never forget the complexity of the task and, beyond quick criticism, acknowledge the author's extraordinary spirit of enterprise.

Writing "about" a book is like "describing tasty food to a hungry man", instead of offering him a good meal. I invite you to read this book and form your own opinion, since ultimately the proof of the pudding is in the eating.

Prof. Dr. H. C. Peter Terness,
Institute of Immunology,
University of Heidelberg,
Heidelberg, Germany

PREFACE

"The pull of the present".
It is extremely difficult to model early earth's
prevalent conditions of early life because we are
so accustomed to the situation from today
William Schopf[*]

For over eleven years since the original publication, *Immune Crossover, the two faces of immunity* deserves further revision. It may steer more overtly to the apparent biological bias to the benefit of groups rather than individuals. The more experienced reader may bypass the section entitled 'Background', which is intended to provide some interdisciplinary tools and conventional ideas about immunity.

Facts and myths regarding immune centrism are still *leitmotiv*. This version accentuates even more the notion of interdependence in the immune sphere. We also continue our attempt to expand it beyond conventional body limits. Focusing on relationships in the context of external environments that we carry or had carried with us also raises the question of interspecies interdependence, especially with regard to microbial flora.

As we discussed, nerve and immune systems share important functions, memory and response tuning amongst them. Although overshadowed by the enormous advantage that communication via light and sound waves has with regard to putative immune interaction, which is supposed to depend at least in part on polluted body contact. Now carelessly discarded, a past role for the 'lost environment' could have been protection against common winter epidemics that increasingly affect modern society.

Shanty towns, rather than being seen in a dismissive context, ought to be seen from a more realistic perspective reminiscent of historical urban over-

---

[*] Adapted from *Scientific American*, September 2005

crowding. This is of particular interest, because in previous editions we may have relied too heavily on a more traditional evolutionary approach which could perhaps justify the term 'time straitjacket'. Now we look at other ways to adapt, even to survive; albeit mostly at the expense of generational transmission, which in fact is the point of Darwin's theory.

However, it is the fear that our evolution may not be able to keep pace with the ongoing transformation that has already caused massive decimation of species and may well be already involved in destroying harmony inside our body. Unaware of moral principles and deeds, immunity can no longer be seen as a default guardian.

Finally, we return to our hypothesis that, beyond body limits, the immune system changes face, that is, configuration and function.

GENES AND 'MILIEU'

FIGURE 1 AFTER SUCH AN ETERNITY, DO YOU THINK THAT MY NEW OUTFIT SUITS ME?

INTRODUCTION

Everything in life needs balance. The outcome of any extremist position, including one-sided scientific ones, is dubious at best; crowded living conditions are clearly such an extreme and, at present, people will be antagonistic towards them. It is not inconceivable, however, that among all the drawbacks of overcrowding, something positive could be found. This book will focus on ideas about the adaptation of immunity in the context of ecology[1] and about possible ways to further extend apparent boundaries.

Recent insights into the new discipline of psychoneuroendoimmunology have led research efforts to concentrate on immune repercussions; or, to put it in another way, mutual immune influences through indirect means – notably via the psyche. Stress, in particular, is being examined from this point of view. There is also the possibility of 'placebo' links (*see* Glossary and page 21).

The fear that there may still exist a real possibility of a sudden global outbreak of plague was comprehensively documented by Laurie Garrett in her book, *The Coming Plague*[2]. The present book focuses on certain immune defence capabilities in connection with the perceived risks. Nevertheless, the situation is not without hope – indeed, it may be appropriate to think in terms of the analogy of a glass of water being half-full instead of half-empty.

HALF-EMPTY?   **HALF-FULL?**

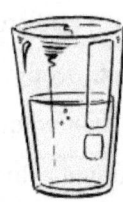
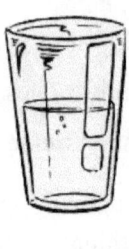

FIGURE 2   HALF-FULL OR HALF-EMPTY?

Humankind takes part in the food chain like any other species and has paid a high price to evolve, the standards of nature being, in general, extremely wasteful. Think, for example, of the ratio of spermatozoa to fertilized ova. Nevertheless, overall, human life appears to have evolved in apparent equilibrium with the environment. To have thrived for so long in hostile surroundings is an astonishing achievement. Whether and, if so, how far, time may be a factor to be accounted for either favouring or hindering evolutionary adaptation of molecular and cellular defence mechanisms is relevant. The conversion from rural dwelling to the 'megacity' also provides evidence of a superb adaptability; nevertheless, an ever-present sword of Damocles hangs over us.

Immune mechanisms are by no means invariant, and a response against a specific intruder varies between individuals. It depends, of course, on the genes encoded by the MHC (*see* Glossary) as a main determinant as to which epitopes are selected to react. In the course of life, the immune network or, more accurately, the integrated defence system (IDS), develops an astonishing flexibility in dealing with threats – but does this imply that it is totally confined within the body; or is reciprocal interaction with other individuals a possibility? Here we attempt to assess aspects where boundaries blur.

Similarly, a distinction should be made between people living in spacious surroundings and those forced to share insufficient space. Unhygienic overcrowding is associated with a multitude of adverse factors, such as pathogenic attacks. Although often negligible and sporadic, they may be massive and dangerous. The process of defence *per se* may be detrimental to the host but immune-defence stimulatory effects are also a possibility. Reacting in concert may perhaps yield better results than individually – a line of inquiry that deserves to be taken seriously – as shared immune information decision-making mechanisms may perhaps link to sophisticated collective immune strategies. Interspecies co-operation should not be underestimated either. By concentrating on the notion of a self-contained immune system[3], totally inside the body, perhaps too little attention has been paid to other approaches.

The idea that antibiotics have almost reached the end of their useful life is already popular; antibiotics are perceived as victims of their own success. For the moment, fears about them seem to be justified, though it is legitimate to ask whether most therapeutic antimicrobial agents have already been discovered. Palaeontologists discovered that evolution – of, for example, humanoid skulls or dinosaur snouts – seems to have proceeded at a leisurely pace, as a result of which it was never imagined that today's antibiotics and insecticides would become so quickly obsolete. However, the fast acquisition of drug resistance is conditioned by the huge numbers of microbes and their ephemeral intermitotic period. Furthermore, it has been ascertained that, at least in the fruit fly, resistance to cyclodiene preceded the introduction of the insecticide. Acquired defence molecules may be spread on to other individuals, even those of different species. Here it must be stressed again that, most critically, drug resistance caused by a single point mutation can be quickly spread through modern transportation facilities. In classic Darwinian fashion, the more antibiotics and insecticides are used, the faster the defence strategy of microbes will evolve. Those that adapt will survive and proliferate.

During its overall life cycle, an organism is exposed to a vast variety of pathogens in its diverse stages of development, mutation and evolution. For better or for worse, the more congested people's habitat becomes, the more the microflora settled at body surfaces, especially orifices, may expand and activate. Such '*milieu*' may also attract germs from the surroundings. Microbes influence each other in different ways: they may repel each other or aggregate in clumps, etc. Interactive processes evolve in the habitat among species – particularly in the relatively stable local flora. Most of the species are optional saprophytes, but opportunism is important in species' ecology. Danger does not necessarily come from afar. Thus, diverse aspects should be considered, not least the long-standing microflora that, in the gut, tends to perpetuate over generations but, on the other hand, displays surprising malleability.

It is an unfortunate fact that the whole of reality cannot be captured at once, and often research projects need to be framed by hypotheses. It should be remembered that in the early days of conventional immune research, advances were made through speculation. Two opposing letters appeared in the London *Sunday Telegraph* of 10 August 1997. One letter claimed that 'at last the truth about the "chemical soup" in which we live is getting some publicity'; the other (acting as Devil's advocate) cited Bruce Ames[4], who has stated that 'roughly half the natural and synthetic chemicals ever tested are carcinogenic and, interestingly, 99.9 per cent of pesticides that humans ingest are natural'. In other words, the quantities of natural pesticides dwarf those of synthetics, and yet neither is poisonous in everyday life. Our body's defences against chemicals are, on the whole, general rather than specific. Ames denies that synthetics have any special potential to cause harm, as environmentalists often claim, adducing evidence from America that shows that the quarter of the population eating the least fruit and vegetables (which are said to be laced with pesticide residue) has double the incidence of most types of cancer compared with the quarter eating the most fruit and vegetables.

Every expression of nature reveals interdependence, or, rather, interconnection. Although in the past some gifted researchers have made serious attempts to take a general approach, in past decades this has not been the norm. The present progression of knowledge requires joint efforts to bridge disciplines. So far, there are no proven circumstances in which external immune relationships are relevant, but every human achievement of insight has a pre- and a post- …

BACKGROUND♥

♥ Contains some 'tools' intended to facilitate readers in understanding the subsequent theme.

IMMUNE MEMES

When, in the 1960s, I collaborated with Ludwig Heilmeyer's and Karl-Georg von Boroviczény's initiative for consensus in hematology[5], I learned the importance of global understanding. Common terms with universal acceptance, when clearly defined, are of paramount importance. The ever-increasing body of data not only burdens our mind but also has added to the complexity of the IDS. Considering that cognition and responses of both systems evolve *circa* in parallel, it is not surprising that human physiology also takes advantage of memes as tools to simplify routine, as well as to develop further in order to effect any change as needed for its continuing adaptation (*see* Glossary).

FIGURE 3 SELF-EVIDENCE IN TODAY'S IMMUNOLOGY

INTERDISCIPLINARY CONVERGENCE

A unique characteristic of life is that it is an organized system capable of creating more order from less order. Although this may appear to contradict the second law of thermodynamics, this is not, in fact, the case. The second law applies only to a closed system, so an open system can interact with the environment, exchanging energy, matter and entropy (*see* Glossary) in the process.

The first real attempt to understand how global behaviour might differ from local behaviour came from mathematicians' multidimensional topology. Later on, emergence was proposed as a distinctive feature of the so-called complex adaptive systems (see next section). We tend to focus on small population clusters that are more susceptible to chance fluctuations. In population dynamics, aggressors and hosts alternately gain the upper hand. Through the development by hosts of strong defences, pathogen numbers may be forced to decline, while lowered stimulation by an aggressor may cause immune alertness to diminish[6]. The process makes for a collective homeostasis and flexible immune regulation – a possible mechanism to attenuate an individual's over-extended or deficient response.

One basic fact is that a residual so-called 'free space' in a crowd should not simply be dismissed as empty; it fully deserves a careful scrutiny that includes its immunological aspects. It is quite understandable that many of the notions described depend on recent insights into what traditionally used to be called 'exact sciences'.

COMPLEXITY THEORY

The outside world is too small, too evident and too real for the wealth that fits in an individual.

Franz Kafka

Physics' quantum revolution at the beginning of the 20th century triggered a paradigm shift throughout science – even certain aspects of the linear approach were revealed to be too simplistic and incomplete. In order to overcome limitations, in the last century scientists from different disciplines moved towards shaping a new theoretical framework that allowed them to get deeper insights into many so-far poorly understood phenomena. In doing so, they developed out of the deterministic view to give pace to a new science of complexity[7].

In a sense, all biological systems are complex. Although a precise definition is lacking, complex systems are a subset of dynamical systems that show some distinctive qualities as they are usually open systems, far from thermodynamic equilibrium and displaying strong non-linear relationships that include complicated feedback loops. Perhaps the most interesting feature of these systems is the ability to self-organize and undergo adaptation in the face of a continuously changing environment.

The archaic belief in spontaneous generation, already attributed to the Egyptians, could perhaps have been an antecedent of the evolving awareness of self-organization. Scientifically, self-organization was first recognized by physicists. They showed that, under certain circumstances, order in a given system can autonomously arise (i.e. entropy can decrease) out of local interactions. This global pattern of behaviour is a property known as 'emergence'. To be in line with the second law of thermodynamics, in these so-called *'dissipative structures'* (Prigogine), such decline in local entropy must go together with an increment of environmental disorder. The self-organization phenomenon is found in the most diverse domains. Perhaps most familiar to the reader could be an ecosystem in which complicated relationships of competence and cooperation at local level give rise to a dynamically equilibrated macroscopic behaviour.

Extrapolating the idea of the crowded habitat context, where the distribution of particles also shows a kind of structure that could influence the hosts' defence, would not be so far-fetched, no matter how counterintuitive it could appear at first sight. Could such ideas be related to the way in which people confined in foul prison cells confronted better than expected the host of adversities afflicting them? Creating local order amidst the surrounding chaos could perhaps provide clues to our subject.

Rather than classic determinism …

Determinism began as a philosophical belief in Ancient Greece. Later on, the idea that every action is the result of its preceding actions led to Newton's[*] essentially deterministic laws that were able to predict systems very accurately: 'Everything that occurred would be based entirely on what happened immediately before'.

---

[*] Isaac Newton [1643–1727].

IT'S THE CHAOS THEORY

Would you enjoy listening to a single-note concert (Figure 4)? By allowing all the instruments to interact, chaos may result. But it may be an enjoyable chaos (Figure 5).

FIGURE 4 THE SINGLE-NOTE CONCERT

Like clouds staging unpredictable spectacles, much has to be learned about the spread of disease. What was thought to be an obvious process is not as simple as might superficially appear. Because many epidemiologists and bacteriologists are biased towards studying microbes, work on other factors involved tends to be neglected. Chaos poses problems that defy customary science[8]. To the enthusiast, it shares highlights with relativity and quantum mechanics. As a physicist once put it: 'Relativity eliminated the Newtonian [1642–1727] illusion of absolute space and time, Max Planck's [1858–1947] quantum theory overcame the paradigm of a controllable measurement process, while Laplacian's [1749–1827] deterministic predictability was abolished by chaos.' Such revolutionary intuition could be applied to nature as a whole, questioning everything we perceive. Or could this just be heresy? Perhaps not. By turning from the Establishment's natural tendency to assume it is on the right track, the chance to broaden the scope to ask how order arises (spontaneously?) out of chaos appears more favourable. Furthermore, the random relationship between what goes into the system and what comes out points to interference.

FIGURE 5 ENJOYABLE CHAOS

Through the work of many research centres, an overall chaos theory is emerging. Even the simplest things require re-evaluation, and, by considering the whole, you might discover a kind of order that chaos theorists call 'determinism', be it waves in the ocean or music, to a certain extent opposed to the reductionist belief that used to find the truth in the particle itself. Trying to decipher the particularities of behaviours, for instance, from star clustering, economy, micro-organisms, atoms, as well as from the IDS, chaos theorists may perhaps reduce the growing tendency towards specialization.

Current understanding postulates sensitive dependence on initial conditions, symbolized by the well-known 'butterfly effect'[*]. Tiny errors and uncertainties, multiplying through chains of turbulence, can in theory lead to this phenomenon, rather like the changes engendered by giving an extra shuffle to a pack of cards before dealing. The butterfly effect is no accident; it is to be expected and is rooted in ancient folklore[9]:

> For want of a nail, the shoe was lost
> For want of a shoe, the horse was lost
> For want of a horse, the rider was lost
> For want of a rider, the battle was lost
> For want of a battle, a kingdom was lost.

For a long time it has been taken for granted that any whole made up of many smaller elements may change. When a chain of events is triggered by a moment of crisis, sensitive dependence on initial conditions becomes significant – it is an inescapable consequence of the way small-scale factors interact with large-scale factors. This would not be the case if the immune network could be reduced to a few understandable, deterministic laws, ensuring predictable long-term behaviour. Inasmuch as immunity depends on many distinct ingredients and irregular external influences, it will remain as an unstable nonlinear system. While linear equations respond to closed formulas – being then easily solvable and understandable – nonlinear relationships are, excepting a few special cases, very hard to decipher. Nevertheless, it is under these kinds of relationships that the most interesting and rich behaviours take place. In the immune network – with lots of known and lots of guessed factors – it is

---

[*] By flapping wings in Beijing, a butterfly could unleash a hurricane in the Caribbean.

clear that divergent signalling by messengers may induce capricious 'out of control' interactions.

Lorenz[*] found unpredictability, but he also found structure amid seemingly random patterns of behaviour. An underlying order, therefore, has to be suspected. The disordered behaviour may act as a creative process. By generating complexity, rigidly organized patterns convert to unstable, finite patterns to infinite. One could say that chaos has become not just a theory but also a method.

In mathematics, fractals (fractus = broken) are any of a class of complex geometric shapes that commonly exhibit the property of self-similarity. As distinct from conventional Euclidean geometry (square, circle, sphere, etc), fractals are self-similar objects whose parts resemble the whole, giving rise to a new system of geometry with impact also on physical chemistry, physiology and fluid mechanics. One key characteristic of a fractal is a mathematical parameter called fractal dimension, which remains the same despite being magnified or if the angle of view varies. Fractal dimension reveals precisely the nuances of the shape and complexity of a given non-Euclidean figure. Fractal geometry, with its concepts of self-similarity and non-integer dimensionality, is applied in statistical mechanics, notably when dealing with physical systems consisting of seemingly random features, e.g. fractal simulations to plot the distribution of galaxy clusters throughout the universe or to study problems related to fluid turbulence.

All in all, on the basis of available qualitative and quantitative data, we are far from being able to elucidate the IDS's capability.

On the other hand, awareness of biology's tendency to re-establish equilibrium is essential. To scrutinise pathogens as they move around the observed hosts as expected does not provide a whole understanding of the fate of mil-

---

[*] Edward Norton Lorenz [1917–2008].

lions of people. To discover harmony in omnipresent randomness and complexity you have to be able to identify obscure patterns transcending time and space. Why should we abandon the many trails rescued from the past and from isolated, almost unknown human communities that still live scattered around the globe?

Laboratory observation requires us to reduce the number of variables to a minimum; thus, it cannot compete with complexities like, for example, flow transition from smooth to turbulent. The contrast is even more impressive in the context of a sprawling 'megacity', in which the population relies on multiple air, land and sea connections. Two almost identical outbreaks of disease might develop along different lines, and a minimal error may prove to be catastrophic, which is in line with Lorenz's postulation and may suggest that long-range forecasting of epidemics may not be possible. Biometeorology and meteoropathology have been observed since ancient times and examples can be cited from the Old Testament and authors in antiquity. Although they provide a different view of the remarkable tendency of living things to maintain homeostasis, findings that climate does not exert a major influence on the progress of epidemics come as a surprise. It is also a fact that people are divided almost equally between those who are very sensitive to minimal micro-physicochemical and macro-physicochemical environmental changes and those who are apparently invulnerable to significant atmospheric variations, geomagnetic storms, etc.

FIGURE 6 CHAOTIC DETERMINISM

Microparticles - Noise

Picking out From a Supposedly Unrelated Pool

The foundations for immune communication so far have proven elusive. The gap between a possible external projection and the well outlined network inside the body is still insurmountable. It is to assume that this doesn't exclusively depend on kaleidoscopic molecular developments; we are thus required to upgrade complexity. In the context, we choose microparticles on one hand and noise on the other. The former provide another important aspect to consider. They may act as mediators for the transfer of receptors and other relevant immune molecules between distant cells. They may represent the medium we are looking for in a model of interhost communication[97].

Nowadays the heterogeneous population of the so-called microparticles, mostly sub-cellular membrane coated vesicles, sized 0.1 μm to 1 μm in diameter, attracts the interest of the scientific establishment. Initially, their release had been attributed to platelets, although virtually any cell type is to be accounted for. Accordingly, both T and B, also endothelial and epithelial may contribute, the same as monocytes, neutrophils, etc.

Living beings have evolved amidst ubiquitous noise. We are now aware that, in *sensu stricto*, noise is closer to a nondeterministic landscape and, as such, it cannot simply be identified by wave lengths. To complement we invite the reader to focus on stochastic resonance. On the other hand, waves rather belong to a deterministic scenario. Like any biological resource, an optimal adaptation, 'eunoise', will be limited by both shortfall and excess. It is to remind that, by affecting the microbial world and ours, for good or for bad noise can become most significant. It also means that the immunological sphere will be involved[46, 47].

More about uncertainty

In the case of disease, statements of fact are obviously required, but they do not always provide secure forecasts of future trends. The spread and regression of flu are as complicated as any atmospheric process. Both cases have periodic components, such as the well-established tendency for a flu epidemic to occur during the cold season, although it is not impossible for winter to be warmer than summer. The problem is that prognosis depends on unpredictable factors which seldom have much weight individually – but, like a bomb being detonated, a tiny incident can trigger off a disastrous plague. Hope not only of understanding the initial conditions, but of extrapolating from them to make predictions for the future, is easily foiled: superficially simple initial conditions may lead to epidemics through intricate interactions. Epidemiology is on the verge of a big change: statistics are well established and better than nothing – indeed, many lives have already been saved through the application of statistical analysis. However, to improve our knowledge of the spread of disease would require a vast network of sensors to provide all the necessary data. Furthermore, the computers upon which we now place such reliance can malfunction with disastrous consequences. The question is: can we rely on anything?

Here we are inspired by the idea proffered by Nathan Wolfe[10], who attempts to anticipate a pandemic even before vaccination becomes an option. In other words, human vaccination would come too late to prevent a pandemia. Wolfe's contagious enthusiasm is snowballing to a considerable research effort amidst rainforests inhabited by simians. Because of their parenthood, chimpanzees and orangutans especially are, of course, very similar to humans. Thus, simian viruses easily adapt to the human genomes and numbers of diseases are originated by the adaptation of diverse animal parasites to the human milieu. The target is to harness simian viruses in the rainforest habitat and to annul them before they are able to infect humans. A significant number of strains has been identified in the simian habitat so far and the outlook for finding more in the near future seems good.

Chris Turney[11] was aghast at learning that a poll in the US revealed that around 44% of the surveyed sample of people adhered to creationism. Creationism propounds, among other things, that 'Time' *per se* is not much older than 6,000 years. He considers the last 6,000 years as a particularly benign period and catastrophes similar to those which happened in prehistory are already incubating. 'Global warming' and other conditioning phenomena encouraged the British scientist to offer a rebuttal[12].

STRESS, PLACEBO* AND OTHER EFFECTS

As long as network dynamics in the process of health wait to be deciphered, a proper elucidation of placebo mechanisms seems questionable.

Besides the complex immune cell/molecule repertoire, obviously, hormones, nerve sources, etc., also make their contribution to the IDS[13] and not all of them necessarily derive from the host itself. Certain functions of the IDS may get support from 'linked environments', mainly in/from the lumen of the digestive tract, but also from outward skin. The latter mainly evolves under heaped living conditions.

Interactive biological signalling has been demonstrated inside the individual, yet precise causal links between psychological stress, IDS's function and disease await confirmation. The brain's limbic system (*see* Glossary) apparently assumes a central role in combining multi-systemic actions. The holistic psychoneuroendoimmunological approach[14] may help to explore constitutive feedback loops for determining the dynamic behaviour of the system. An excess of stress may affect the interplay between immunogenic pathogens and neurohumoral feedback. Positive or negative stress, depending on whether its condition is mild, intense, acute or chronic, can enhance or suppress immune responses[15]. It has long been a commonly held belief that social support – for example, the presence of a close companion – can lessen susceptibility to disease by dampening anxiety response[16]. Individuals with a strong social support network exhibit less morbidity/mortality[17], although this may be difficult to measure functionally. More objectively, a similar tendency could be shown in nonhuman primates[18]. In juvenile rhesus macaques, Gust *et al.*[19]

---

* Placebo harks back to the 116th Psalm in the Hebrew Bible. Through a number of translating errors, the Vulgate Latin version came to contain the word 'placebo' (I shall please). Over the centuries, the term placebo was applied to the Vespers of the Dead, derisively to servile flatterers and toadies, and to laments sung at funerals by professional mourners.

confirmed that the presence or absence of social partners modulates response to stress in the immune sphere. Lone subjects revealed a decline in helper T cells (*see* Glossary) and an increase in cortisol levels during environmental adaptation. Recent awareness of physiological changes related to eustress is of special interest. By observing mice in the basal state and in response to exercise stress, Kingston and Hoffmann-Goetz[20] compared differences in physiological exercise-induced environmental enrichment and 'housing' density, the former preventing negative stress-induced enhancement by a promoter of T cell proliferation[*]. Apparently, pleasant life activity induces eustress, thereby providing direct benefit to immune function[112]; it also acts indirectly, by buffering immunosuppression that distress may cause.

The buffer effect of eustress occurred irrespective of housing density. Eustress as a condition for good health is now widely accepted, and the IDS profits from it by mounting optimal responses[21], quicker, more vigorously and more efficiently. There is also a faster recovery to prestimulus levels (homeostasis).

It is clear that people interact strongly through psychological means when crowded together. Stress may act as a potent trigger and in such subjects it echoes through neurons, hormones and other messengers. Obviously, the IDS becomes involved. Recent experimental evidence may allow a broader reading and, ultimately, a mutual immune influence might result, obviating or supplementing other factors (see more page 106).

Cliché- like prejudice in tandem with complex adaptive responses, allowed a million year yet often miserable survival. Until very recently, humankind's main concern (and that of other species) was to overcome hunger.

----

[*] By concanavalin A.

It might be that crowded living could bolster placebo effects as an indirect means of immune co-ordination. The placebo puzzle is persistent and widespread – for example, menstrual synchronization, or normalization of menstrual cycles triggered by intercourse. Placebos often provide straightforward help, but they have also been associated with all kinds of unpleasant side effects. Even 'placebo addicts', although very uncommon, have been observed. For an appreciation of the phenomenon, it might be apposite to imagine Sigmund Freud, in the 1920s, calming down someone's heterodox suggestion that the alarming mental problems affecting a patient could be corrected simply by regulating a disturbed thyroid function. Without answering our questions, psychoneuroendoimmunology has made us more flexible.

GENETIC INTERLUDE

In its January 2009 issue, the *Scientific American* highlighted Harpending & Hawks'[23] team, which estimated that humans have evolved over the past 10,000 years, as much as 100 times faster than at any other time since the split of the earliest hominid from the ancestors of modern chimpanzees. They attributed the quickening pace to the variety of environments humans moved into and the changes in living conditions brought about by agriculture and cities. It was not farming, *per se*, or the changes in the landscape that conversion of wild habitat to tamed fields brought about, but the often lethal combination of poor sanitation, unusual diet and emerging diseases (from other humans as well as domesticated animals).

Although it may be tangential, the reader may realize the significance of the following inserted paragraph. Recent research strongly backs the often rejected idea to generalize about the male condition. While men are comparatively 'predictable', women are revealed to be genetically more complex than scientists ever imagined – creatures of 'infinite variety', as Shakespeare wrote. A whopping 200 to 300 (15%) genes from the second X chromosome are active at levels greater than it had hitherto been known – not submissive and inert, as previously thought. According to Nicholas Wade[*], due to the edge in gene expression, women are mosaics, one could even say chimeras, in the sense that they are made up of two different kinds of cell. While women are getting the benefit, in men only 45 chromosomes work in full; the 46th is the Y and has only a few genes to operate. After shedding genes for millions of years, the Y chromosome is now falling apart. In 10 million years, or even much sooner, men could disappear altogether. This new awareness may help to explain why the behaviour and traits of men and women are in effect different; they may be hardwired in the brain, in addition to being hormonal or cultural. Women's greater complexity in chromosomes may be the foundation for 'unpredictability'.

---

[*] From the *New York Times*.

TIME AS A STRAITJACKET?

New Insights in Adaptive Shortcuts

Time heals all wounds – but it is also the great destroyer.
Attributed to Jorge Luis Borges.

Commonly, the IDS is believed to be predictable, but when the environment puts our species' survival at risk, unexpected changes cannot entirely be ruled out. Many open questions that pose the 'biological clock' should not be underestimated either.

The classical view of evolution states that this process relies exclusively on random events – mutations, deletions, duplications, etc. – in order to generate genetic variation. Such variation gives pace to the phenotypic change that is then subjected to natural selective pressure. From this perspective, the so-called Darwinian evolution in principle takes place on time scales ranging from thousands to hundreds of thousands of years.

Recently, however, the traditional laws of evolution have been put under re-evaluation. Lynn Margulis' endosymbiotic theory[24] and John Cairns' famous experiment giving rise to the premise of non-random mutations[25] were among the first attempts to go beyond the classical evolutionary theory. Since then, many mechanisms have been suggested as complementary to natural selection in driving evolution. Of special interest to us is how individuals overcome the limitations imposed by the time straitjacket in order to stay one step ahead of the consequence induced by rapidly shifting environments[26, 27].

Phenotypic plasticity, the expression of alternative phenotypes arising from a single genotype in response to environmental fluctuations, represents a recent paradigm in evolutionary biology[28]. Although intimate mechanisms remain to be elucidated, apparently, the time factor doesn't necessarily have to be at the forefront. A putative role is relative at most. As stated, a polluted population density may be seen as a main source of environmental variability able to trigger phenotypic changes. Among others, we suspect also that tools to improve the level of defence may come to light. By assuming deviation in immune behaviour as a distinctive feature of congested human population centres from the past, phenotypic plasticity is to be believed as pivotal in this transformation. Here, we deal with immune challenges that require a swift adaptation. Thus, besides the dangers of plague, famine, etc., on the other side of the coin, factors may emerge that made survival possible. Phenotypic plasticity may, in fact, introduce radical modifications, although these will not be transmitted to a next generation. Furthermore, once the causal factors run out, the question is whether the particular immune response mechanism does revert. Both ways may act independently or within a sort of tweaking-back-and-forth cycle.

Symbiosis is a process that could provide an organism with a shorter evolutionary path by conferring on it some features it has not had to evolve on its own. Furthermore, symbiosis can be seen as a mechanism by which an organism can enhance its robustness. Robustness is a property that allows a system to maintain its functions despite external and internal perturbations. The theory of biological robustness argues that it fosters evolvability and that selection tends to favour individuals with robust traits[29].

In this context much has to be said about the profile of the IDS. It is thought to have been shaped over eons – overcoming many environment-conditioned ups and downs. Development is not supposed to have been just smooth – the astonishing potential of the IDS to adapt remains to be elucidated. Intuitively at least, the time factor is a main player.

Before pursuing with our main target, some examples that reveal a pace of adaptation that is hard to believe and that may perhaps conceal helpful clues are worth a mention:

Think about the sudden awakening of trillions of cicadas, an event that recurs every 17 years, when the onslaught of the cacophonous beasts literally covers a vast expanse of the US.

Another perspective provides the sudden decline of flourishing civilizations. Archaeologists are now committed to unfolding the dynamics of the dramatic end of most of the significant Mayan cities. It is estimated that one of the most ancient cities, *circa* six millennia ago, had a population of about 100,000 people. It is highly suggestive that, in the first instance, a formidable construction demand may have accounted for their downfall. Denuding rainforests by deforestation would have induced devastating droughts, triggering famine and warfare.

Resuming the theme, two seemingly opposite concepts have so far been mentioned. On one hand, rather than being advantageous, most mutations are likely to be neutral or harmful to the individual. Therefore, they are a source of perturbation to which an organism must be robust. On the other hand, evolutionary adaptation requires some mutations to induce new traits. Researchers have now found that the so-called *'promiscuous proteins'* may hold the key to solving this apparent contradiction. These proteins have the capacity to evolve along many different functional lines and are able to perform different functions. When exposed to stress conditions, they have been shown to maintain their main function unaltered, but at the same time to change drastically the level of promiscuous activities – a series of secondary activities, mostly unrelated to protein's main task. What is more, if the selective pressure is strong enough, it could lead to divergent protein lines. Perhaps as a challenge for proteomics, 'promiscuous proteins' could be seen as an intermediate stage in the evolutionary pathway[30].

Towards its present status, species' evolution, obviously implying the IDS, underwent distinct phases that conventionally are thought to depend (among others) on the time factor. From this point of view, by lacking the chronon for the IDS evolution (*see* Glossary), we are not in a position to reconcile our prolonged past and the accelerated globalization presently under way.

In brief, we are now conscious that the idea of a '*time straitjacket*' has been overshadowed by new options and expectancies. They may explain more convincingly the long human odyssey.

Communicativity or Non-Communicativity?

The astonishing creativity and freedom of thought of infants have been acknowledged for a long time. To investigate their potential to solve certain complex problems in science, attempts have been made by selecting highly gifted children as young as three years of age to co-operate. Among the concepts that are prone to confusion is non-communicativity. The non-communicativity of actions has implications ranging from the trivial to the profound. Alexander the Great [356–323 BC] vented his legendary impatience in untying knots by slicing asunder the Gordian knot. Looking at systems by examining isolated molecules, and then integrating the results, has no future so far. Moreover, non-communicativity obstructs a proper understanding of the possibilities suggested by considering collective immune behaviour.

ABOUT ECOLOGY

Increasingly, the view is shared that the globe as a whole functions as a network and in this context it may be considered as an organism[31]. Ecology implies a reciprocal relationship between things. If this is a global property, it cannot be assumed that the immune system will be an exception. As a consequence, subjects manipulate the immune environment which, on the other hand, manipulates them. This raises the question: can the environment integrate immune interaction through ecological means? Over time, so we speculate, researchers will evolve a malleable public immune-homeostasis, preventing people from over- or under-reacting.

Current risk factors are, of course, many. Here we point again to deforestation that nowadays is global. Such recurrence should have been prevented, as it also happened in the late Middle Ages, almost coincident with what was known as the Black Death, which killed about one quarter of the population of Europe; in some places three-quarters of the population were wiped out.

Global communication allows us to simultaneously alert billions of people to the dangers from meteorology, overpopulation, consumption excess, etc. – the good news is that, increasingly, people tend towards a healthy response – maybe improving the outlook for all.

HISTORICAL THOUGHTS RELATED TO INTERACTIONS

For Aristotle [384–322 BC], physical motion was not a quantity or a force, but a kind of 'change'. Modern physics stresses that many kinds of observations only become clear if subliminal effects are considered. So the idea that overcrowding could facilitate immune contacts has to be considered in terms of point interactions rather than in terms of integration.

Soul - Mind - Body

Plato [427–347 BC] believed that the soul was fundamentally distinct from the body, although often it was affected by its association with the body. In spite of being simple, rather than composite, and thus not liable to dissolution, as were material things, the soul had the potential to rule the body. Plato denied the Pythagorean [570–497 BC] view of the soul as an 'attunement' (*see* Glossary) of the body. For Aristotle the soul was neither material nor immaterial, but an abstraction. In plants this was concerned solely with nutrition and reproduction, but in animals sensation and independent movement were added, while in men the soul's crowning achievement was rational activity, although this was also thought to be under outside influence.

From René Descartes [1596–1650] onward, more was heard of the mind than of the soul – the individual's mind especially. Descartes held the view that mind and body are capable of affecting each other causally, so that what happens in the body can produce effects in the mind, and vice versa. He stated his belief that one's soul is not to be found in the body like the captain in a ship, but is more intimately bound up with it. We may remember the famous *cogito argument,* 'his only certainty is the existence of his mind, for the very act of doubting was itself mental'. The Cartesian account of mind and body had many critics, even in Descartes' own day; when, for example, Thomas Hobbes [1588–1679] argued that nothing existed except matter in motion[32].

30

Baruch Spinoza, in fact Despinoza, [1632-1677]. contributed to the slow process of detachment of philosophy from theology; a single entity, God or Nature, possessed of infinite attributes, of which the mental and the material alone are only known to Man[33]. Its manifestations were the result solely of its Nature, and arbitrary will neither did nor could play any role in them. Whatever manifested itself under one attribute had its counterpart in all others. John Locke [1632–1704] thought that 'sensible' ideas of size, shape, position and motion or rest resemble material objects as they are, and this is in agreement with the scientific view of modern science, which is based on the mathematical focus on primary properties. Locke's ideas were further evolved by David Hume [1711–1776], who put mathematical objectives above those founded on emotions. Hume's distinction between fact and value issued from the elevation of the mathematical objective over the emotional and the subjective; minds and bodies alike were nothing but 'bundles of perceptions', between which interactions were always possible[34]. Action and cognition became less dogmatic when Georg Wilhelm Friedrich Hegel [1770–1831] favoured a closer inspection of the facts, Great impact had Immanuel Kant [1724-1804] by championing a renewed epistemology. Nevertheless, even at the beginning of the twentieth century, the belief that mind and matter are constructed out of neutral monads still reveals the influence of the early Cartesians. Nowadays machines have also become involved. They do not, of course, literally think, but they are revealed to be capable of astonishing performances. They even have an analogue of consciousness in the sensitivity they show to external stimuli.

Like any logical process, immunity and its dynamics are subject to modification – in the context of any given milieu and time. We know very little about the entire immunological orchestration in historical terms, but it would be naïve to pass over possible dynamics. We must keep alert because the characteristics of plague are supposed to be subject to equivalent changes, although a strict parallelism is out of the question. Notwithstanding impressive immune cognition, it does not seem to make sense to attribute this to a mechanical process.

Self

Stress on the 'self' or 'ego' is behind Hegel's developmental idealism, when he envisioned a World Soul coming to consciousness. John-Paul Sartre [1905–1980] argued that each individual chooses to come into being out of nothingness. He upheld the Cartesian position that the self is conscious by denying the unconscious self proposed by Sigmund Freud [1856–1939]. For Gilbert Ryle [1900–1976], the mind (ghost) is simply the intelligent behaviour of the body.

Interaction

Interaction is a universal phenomenon which has been shown to occur even at the cosmological level among certain pairs of galaxies (with spectacular results for astronomers). Although interaction is deeply rooted in biology, retrieval may be difficult. In this respect, defence mechanisms also have to be considered. To place them exclusively inside the body would be a contradiction. Therefore, population defence mechanisms have to be seen accordingly. Relating the IDS with organisms as a whole appears to be in line with ideas already outlined by Pythagoreans and Cartesians. Levin and Solomon related the evolutionary biological entity we call body to a dynamic equilibrium balancing between health and disease on the one hand and progressive socialization on the other. While medical thinking sequentially focused on anatomy, physiology and chemistry, now there is a trend towards a more holistic perspective[35, 36].

Here we must call the attention of the reader to the astonishing achievements of social insects such as termites, ants and honey bees. Recent insight into the reciprocal functions is seen increasingly as a form of superorganism. It obviously raises the question as to whether the immune function does contribute to such society. As a point of fact, such regulatory function has been observed in the case of termites[37].

Biotic Interactions

No species lives in isolation from other species. The universal fight for survival, that includes the whole food chain, ensures a constant struggle; that plants and animals are attacked by parasites is part of it. Biotic interactions may occur between members of the same species (intra-specific) or between two or more species (inter-specific). Different species generally interact with each other in competition, but sometimes to their mutual benefit (symbiosis). Besides positive or negative effects upon the participants, the members of an interaction can remain apparently unaffected. Some interactions are temporary, casual and of minor importance, whereas others are permanent, vital and of major significance. The diverse interactions, which evidently include defence mechanisms, are often highly specialized and complicated, involving adaptive changes in structure, function, behaviour and ecology[38]. Biotic interactions are significant not only because they influence individual species but also because they constitute the principal stabilizing, connective linkages among the various species contained in a community. In a general sense, species in a biological community maintain a relative harmony because of biotic interactions, despite individual gains and losses. As a consequence, the community as a whole persists, with all individuals contributing in some way.

Mutualism is an association of two species resulting in mutual benefit or gain. Facultative mutualism, or proto-cooperation, refers to an association in which one member can survive in the absence of the other. Commensalism means that one member benefits while the other remains unaffected.

As will be discussed in the context of the astonishing ecosystem settled in the colon, mutualistic co-existence strongly suggests that many microbes are needed for continued human existence, yet the impact of micro-organisms on

33

homeostatic and physiologic processes is still incompletely realized. Better understanding of these interactions could lead to strategies to improve the well-being of humankind. In this context we also have to consider the pros and cons of the means by which outside interactions take place. (see 'Skull Boxes and Skin Bags')

OUR PAST

"But also may it please you to re-member that it is an anvil that has worn out many hammers".

Theodore Beza

History repeatedly refers to cities that went through a prolonged siege. Only a small percentage of the population would normally live in the congested forti-fied cities. Even under normal conditions, the space available for permanent residents was insufficient – indeed, to this day there still remain many places in which an average of six to ten people sharing a room is the norm. To pre-vent smuggling supplies, the inexorable progression of Titus' legions induced neighbourhood's flight.

FIGURE 7 REGARDLESS OF THE INSIDE ENVIRONMENT, THE WALL IS THEIR LAST HOPE.

Although no exact figures are available, there is no doubt that, during the Roman siege of Jerusalem, crowding became unsustainable by our standards♦. How people inside the city managed remains a mystery. It is hard to imagine death all around and people in close confinement submerged in filth, piling up excrement and rotting remains, and deprived of water. Regardless of the consequences, starvation forced them to consume whatever was available, exposing them to havoc from toxic remains, pathogens or saprophyte attack. But, in spite of everything, defence did not disintegrate completely. The fighting was often prolonged.

How did the individual's immune response cope? Certainly the particular environment required special adaptation. It is possible that, from an immunologic point of view, sticking together turned out to be advantageous. One must infer that those who had no opportunity to co-adapt could not match the resistance capacity acquired by the besieged, infants and old people among them. Any clear understanding of the diverse interwoven immune dynamics that evolved will remain elusive for the time being.

FIGURE 8 THE TWO FACES OF IMMUNITY

♦ To make things worse, Titus allowed pilgrims to enter Jerusalem to celebrate Passover, and then refused to let them leave.

Another fact from our recent past was the introduction of slaves into the workforce in urban communities. It was very hard for them to survive under the inhumane conditions imposed on them, with malnutrition, etc., as the norm. Chronicles tell that they often succumbed to infection, not infrequently with epidemic features. Mishap indeed came in a variety of guises but, for purely practical purposes, in order to get the expected return for their money, the masters had no choice but to feed their 'product' more or less adequately. The same purpose moved them to select it at the prime of life – also the optimal immune age. To mount an appropriate defence, the immunodeficient age groups (both old and young) of the stable population must have been a greater risk. It is conceivable that the main handicap of slaves was their condition as newcomers. In contrast to the stable population, they were not used to the environments of filthy congested towns, hamlets, or even farms.

FIGURE 9 SLAVE SHIPMENTS PROVED TO BE WASTEFUL IN THE SHORT TERM AFTER ARRIVAL

Slaves' immune profile should do better than that of infants and old people, but collective immune streamlining requires adaptation.

History as a Link

Bridging a million-year-long prehistory to our current 'globalization', history has to be seen as a short period. The sharp contrasting periods highlight time as a most intriguing factor in the evolutionary adaptive capability of our IDS. Nomadic life – hunter-gatherers – had been a characteristic of prehistoric man. With the advent of systematic farming, sprawling population centres emerged in a world that was considerably less populated. During the historical period, nomadism continued around the world often lacking any contact with villages or towns. In the cities, particularly when high birth rates surpassed the death toll, people were compressed into a minimum of space. Poor heating may have been a contributing factor. Furthermore, urban settlements increasingly attracted protection-seeking people.

Until quite recently, the lack of hygiene was a common denominator. No doubt urban confinement meant a major change in so far as acquired rules by the IDS game (see more in the *'milieu soup-interface'* section, page 104). In this regard, we suppose that a certain attenuating buffer effect operates in the group. An especially collective behaviour meant that, in spite of hardship, it was people from over-populated urban spots that brought civilization and culture into being. Almost inconceivable to us today is a scenario described in a letter written from Paris by Wolfgang Amadeus Mozart to his father, talking about lost piano revenues. Having been refused, he attributed the lack of money to being driven in streets full of 'shit'. Thus, forced to walk in the cold, he also complained about stiff fingers interfering with his skills in playing the pianoforte.

The fact is that human settlements accounted for the progression towards modernity. Civilization and science have to be seen as being derived from these overcrowded centres. Although we are now aghast, it may also have meant a certain collective immune advantage. The resulting dualism was not only reflected by the evolving changes in the way of life – it is supposed to have

38

had an influence on epidemiological patterns as well, leading perhaps to functional differences in the respective IDS.

Industrialization means a fundamental transformation, significant in the
turning point we are witnessing. It is, of course, fundamental in luring people
even more to concentrate, as being reflected by today's megacities. In other
words, the historical conditions appear to untie even further the gap in a parallelism, as revealed by a row of meaningful neural and immune properties. The
latter's inability to sense waves of light and sound makes a proper appreciation rather difficult at first sight. According to our hypothesis, the resultant
by-products stand for both an explosion of civilization and science and an
implosion, implying the loss of the bud of a collective immunity (Figures 10
and 11).

PREHISTORY
million year trickling

HISTORY
intermediate course

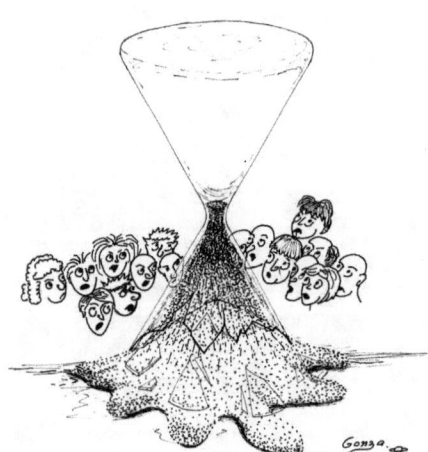

GLOBALIZATION
flooding

FIGURE **10** DYNAMICS OF CHANGE **I**

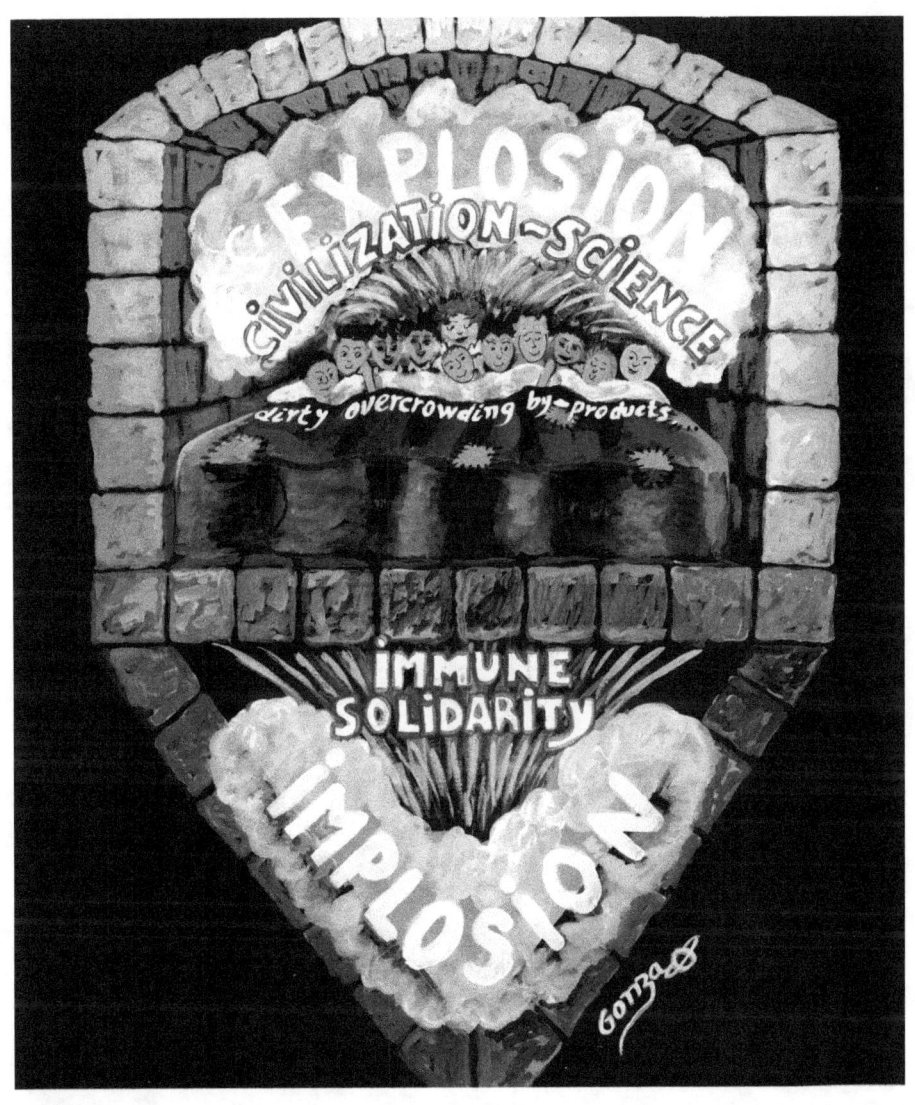

FIGURE 11 DYNAMICS OF CHANGE II

IMMUNE DEFENCE
BALANCE - IMBALANCE

Some Introductory Concepts

All multicellular organisms have highly effective signalling systems enabling cells to communicate. The properties of networks emerge from their organization rather than from the characteristics of their individual components. Signal transduction pathways have been observed in both tissues and body fluids[39]. Another cue reminding us that the subject is in continuous expansion is current awareness of a role played by microparticles in immune signal transduction (*see* Microparticles).

The ubiquity of the IDS network is now widely accepted. In almost every part of the body, by a kind of default, immune cells and messenger molecules, neurotransmitters and hormones coordinate defence – the diversity of protagonists is still increasing. As mentioned, receptor molecules, to a greater or lesser extent, are shared by many cell types. Not all are necessarily membrane fixed; a number of them are moving freely. Immune responses depend on conditions such as target concentration and enhancing and suppressive effects often combine.

Emerging disciplines like psychoneuroendoimmunology (names may differ) try to pinpoint how developed organisms orchestrate the IDS in harmony with other body functions. The fact that many receptor molecules are not necessarily specificity restricted may come in support of the present network approach.

Efficient cellular and humoral responses require large but finite numbers of mature antigen receptor-defined immune cells. To initiate an optimal regulated interplay, antigen presenting cells, mostly dendritic cells (*see* Glossary), respond to a primary stimulus, a process that also requires a 'co-stimulator' molecule (*see* Glossary). Usually, they activate lymphocytes from the type T

43

that may lead to a cell-mediated immune response or a humoral response by B cell synthesized antibodies. Diverse types of lymphocytes and specialized accessory cells are essential. The specific clones in charge of supplying mature lymphocytes not only adjust to homeostasis to prevent overexpansion or depletion of specific lymphocyte sets, but also to provide long-lasting malleable immunity. Nuances in the different elements deserve appraisal. For host *versus* germ reactions, it is not only the quality and spatial location of the epitopes that matter. Often, antigens convert from a basic form to whatever the milieu is able to modify, requiring different means of tailored immune responses. Immune responsiveness against a specific virus varies between individuals. The presentation of an appropriate peptide from the selected epitope depends, among other factors, on genes encoded by the MHC. In other words, the encounter leads to adaptive changes for aggression and defence, respectively. This could be another aspect to consider among possible causes that prevent some sex workers from contracting human immunodeficiency virus infection (HIV)[40]. Immune responsiveness can modify over time; for example, in HIV as a result of antigenic mutation (*see* Glossary), or declining T cells. The IDS, by integrating responses throughout the body, may account for area-specific modalities; for example, in the alimentary tract, respiratory system, or delimiting skin. For healthcare, an overview of effector sites and immune traffic seems of interest.

Instead of working, as hitherto, like mainstream physics that simply considers properties of particles, such as whether they are tiny or huge, a new approach may provide more light from a different direction. Gradually, it may become apparent that standard assumptions may not fit the progress of an epidemic. Minor discrepancies may build up into significant consequences. Last but not least, as will be discussed later, the 'immune confinement dogma' may have contributed to postponing efforts for elucidation.

Prejudice

It does not seem too difficult to introduce the idea of bias as an immune function. A stimulus, provided that it turns out to be customary, will tend to simplify both the cognitive and the reactive arms. By 'learning' to save energy they act in a typical biological way. If this is so, a comparison with the social aspect of prejudice would make sense. In essence, the exclusion of aliens by prehistorical hunter-gatherer clans and modern marauding simian tribes would not make much difference. In both cases, energy saving may be seen in the first place as merely background.

Integrated Defence System (IDS)

Any open-minded approach to the immunological theme has to account for the interplay of the central nervous system (CNS), the endocrine system and the immune system. Interactions come to light, especially under stressful events. Here we recap that, in principle, so-called eustress does not interfere with physiological homeostasis; rather, it is supportive of an active lifestyle and is also required for health recovery. Worrying ranges from commonplace daily hassles to chronic calamities. Chronic psychical or physical stress can be deleterious to immune function. Depending as it does on the complex network of bidirectional signals between the nervous, endocrine and immune systems, the term IDS is well suited (*see* Glossary). Also, the interdisciplinary field of psychoneuroendoimmunology is moving forwards. Furthermore, microbial partners allow us to delegate scores of functions. Because of the extended genetic landscape*, Bäckhed *et al.* regard *Homo sapiens* as an amalgam of many species[41].

---

* Our genome may be exceeded a hundredfold by the genes from the intestinal flora.

45

Inflammation

Calor, rubor, tumour and dolor.

Celsus[*]

When any part of the human body surface gets hurt – by flame, deep freeze, tiger bite or countless pathogenic invaders – the expected response will be an inflammatory reaction. From short-lived onwards, according to the circumstances, inflammation may become chronic. Also, the intensity is variable, the same as the general repercussion – symptoms – the individual may manifest. Severe systemic inflammation can be induced and perpetuated diversely – be it by toxic bacterial products such as endotoxin[42]. In most cases, inflammation means a 'counterattack' to the benefit of the host. The outcome may be uncertain, of course. It is problematic that autoimmune inflammation is also on the rise. As will be discussed later, here the IDS turns aggressive against 'self' – damaging its own cells or tissues. Simply, at least during the optimal segment of life, inflammation has to be seen as a first step in the healing process towards *'restitutio in integrum'* – an overall biologic tendency.

Often, an inflammatory process starts as a highly specific adaptive immune response in tandem with, last but not least, the definitely less specific innate immunity. Typically, to 'fire' the complement system or perhaps the clotting system – enzymes' assembly activate the cascades – so far the trigger gets to threshold (e.g. cell-secreted molecules such as proinflammatory cytokines, antigen-antibody complexes, lipopolysaccharides and other bacteria-derived substances, etc.). Blood vessels and the local extravascular space, nerves inclusive, belong to the inflammatory orchestration.

It may be of comfort for the rising body of patients affected by chronic inflammatory diseases that the so far uncertain outcome of immuno-modulation

---

[*] Roman physician of 1st century AD. According to the *Encyclopaedia Britannica* (1992), as a description it has still not been surpassed.

– for example, high-dose intravenous gammaglobulin infusions (IVIG)♥ –
may finally be realized[43].

Skin

Throughout the organism many immune responses integrate with the involvement of body surfaces – gastrointestinal, respiratory and genitourinary tracts among them – normally maintaining associated microflora in check. Skin, as the apparent front line, also exerts a certain auto-control essential in defence. In spite of playing down the idea of absolute boundaries, skin – together with less tough mucosae of respiratory and gastrointestinal tracts – is more than a mere mechanical barrier. Proof of this is the nightmare of infection resulting from sizeable burns. All cell linings are constantly being renewed from underlying layers. The keratin-producing outer layer of cells efficiently covers the skin surface, and glands cooperate by secreting oleic acid, lysozyme, etc., that can destroy bacteria. In the more developed vertebrates, a great variety of immune elements cooperate. It would nevertheless seem absurd to deny the interactive IDS network the possibility of traversing delimiting epithelial linings. Besides, the noteworthy orifices that offer plenty of opportunity for immune transit remind us that, after delivery, breastfeeding prolongs the protection of the offspring and something equivalent happens in calves via airways.

Via Skin

Penetration via skin, albeit slow, is now widely accepted as an important therapeutic tool. Its popularity is founded mainly on constant, controlled drug penetration – a convincing advantage in comparison to conventional administration by the oral, intramuscular or intravenous routes. Awareness of access

---

♥ Pooled immunoglobulin, mostly the gamma isotype.

through epithelial linings underscores that a whole range of alternatives remain to be explored and the diverse secretions and substances released via body orifices (urine, faeces, sperm, etc.) support the notion that the immune network extends beyond conventional limits. If we focus on possible immune network connections between individuals from the same species, we have to rely on the transfer of molecules and, to a lesser degree, immune cells; body closeness being a precondition.

Senescence: Beware Of Hasty Conclusions

> *"By getting old, a political party, humans alike, will invent a disease to fade away"*...
> Corriere della Sera

> *"An old man made two heavy guys flee. How? He couldn't explain."*

It is universally accepted that advanced age increases vulnerability to all kinds of aggression; the ambiguity (see later) of commensal flora can also be a risk factor. To blame the IDS exclusively is not acceptable – recent reports draw attention to often associated nutritional handicaps. We must not underestimate the fact that, to a major or minor degree, immune homeostasis takes part in physiological and physical deterioration – as do circulatory, nervous and hormonal functions. And the profile diversity of genes also accounts for individual fate. As in the case of mental decay, in which memes tend to endure, what we call immune memes may perhaps come into their own by curbing an ongoing immune paralysis. Whether molecular biology will come to be of help is a matter for current discussion.

About The Functional Window

A boxer at the edge of a KO will not give up, but, with no system, his fighting becomes grotesque.

Conventional wisdom tells us that in biology every function has a threshold and a functional 'window'. Inside the window, immune challenges tend to induce efficient responses[*]. There is even consensus that a certain immuno-modulatory benefit may be expected by stress, providing it remains in the eustress zone. Physiological and even disease repair remain unperceived in most cases; when revealed to us, manifestations are usually vague; but symptoms soaring to startling heights are another possibility. Overreaction, either immune or nervous, may result in harm to the host. An immune response may manifest itself in fever, shivering, pains, etc.; symptoms mainly correlate with pro-inflammatory molecules released in excess. This is in sharp contrast with a collapse by overwhelming aggression – the limit is, precisely, the 'functional window'. Beyond it, any systemic function ceases.

As a medical student, I was taught that if an old man with pneumonic lung consolidation and breathing discomfort fails to manifest fever or other clinical signs related to a possible response by the IDS, he probably will not survive.

I remember an exhausting epidemic of 'flu that raged over the country. All three younger generations of our family were down with the full-blown picture of the disease, but both my 94-year-old mother-in-law and, 400 km away, my 95-year-old mother in Buenos Aires did not catch the scourge. They ought to have been invaded by the virus like everybody else. Despite failing to react, the outcome was uneventful – supposedly, plasma proteins impeded the virus from thriving.

The idea may be applicable in human history:

---

[*] This has to be viewed in the context of transformation as determined by leading and lagging immune processes.

At the dawn of the 19th century, Heinrich Heine[♥] acknowledged French philosophers who helped to open the gates to their great revolution; although, prophetically, he also alerted the French that the Germans were more meticulous: 'Once their time to fight has come, no trace of their superb philosophy will remain. Blindly they will produce death and destruction all around.' Could we say then that historical transformation in France, but not in Germany, evolved a functional window inside?

In essence, every biological phenomenon depends on a particular functional window.

[♥] [1792–1856].

'Milieu Soup'

Substances reach hosts by a combination of dispersal and resolution that is determined by two kinds of processes termed vectorial and stochastic. Vectorial refers to directional dispersal by environmental motion, such as in wind and water currents. Stochastic processes are essentially random forces, the operation of which does not allow a prediction to be made as to where particles will be carried and will settle.

The term *'milieu soup'* has been coined to describe a densely packed interface as, for example, in the digestive tract, where it may take the place of the more popular term 'juice' (gastric or enteric). Here we apply the term to illustrate our idea of an intense particle traffic conditioned by filthy, squalid living. A *'milieu soup'* beyond skin seems justified for a particles pack with a significant organic component – in the first place, we think about immune molecules and also a certain amount of host-derived cells. In this context, a probability of immune interactions, even functional, may be considered. In other words, influence may not be restricted to parasites as it could even affect the counterparts' immune sphere (*see* 'To consider'). It would not be too surprising if such a heterogeneous-formed combination, springboard-like, could imply initial conditions for a self-organizing body belt system. Such systems account for steady modifications, and nobody can deny them a mini-probability of unforeseeable consequences. Although unpredictable, marginal conditions are essentially criteria of a selection from the great variety of possible processes. In a way, they might delimit those which in fact happen. Information *per se* is not totally alien; one could dare to apply the same to the idea of collective immune mechanisms.

Early Evolution of Host Defences

The immune system is inseparable from an individual's integrity and therefore takes its phylogenetical roots from the evolution mirrored by most primitive beings. Single-cell organisms defending themselves against the environment – although they do not have a distinct cell organelle which could be ascribed to the immune system – behave immune-competently insofar as they protect their integrity against the outside world. Higher plants and lower animals possess tissue antigens that prevent simple transplantation of parts, even when derived from individuals of the same species.

Developing specialized defence systems takes part in the evolution. For example, it is known that certain marine invertebrates possess proteins of the alternative pathway of complement, and this in the absence of any specific immune response. Higher up the evolutionary ladder, defence mechanisms increase morphological and functional efficiency. The systemic patterns they deploy are called the immune system or, more appropriately, the integrated defence system (IDS).

The paramount importance of chemical communication in biological systems is now widely appreciated. Messenger molecules coming via body fluids or from the immediate neighbourhood may ligate to cell membrane-bound affine receptor molecules. The elucidation of molecular interaction diversity is still on course. Interconnection is essential, and chemical signals in their widest sense are a universal attribute of life within and between all organisms. It is remarkable that many active molecules from all kinds of species work in similar ways on the basis of only a few essential building blocks that constitute biology. Further, the omnipresent capacity to distinguish 'self' from 'not-self' is impressive. Both individual and shared features of immune response reveal species-efficient tendencies. It could be that evolutionary conservatism has practical implications on the environment's immune properties.

In fact, long before there existed awareness of the immune network, people realized through commonsense that infection could spread through contagion and that in a number of diseases, when recovery is achieved, it means a long-lasting protection. From early times this observation led to the development of biological warfare by catapulting the bodies of victims of pestilence into besieged cities. Thucydides [*c.* 460–400 BC] narrates how, during the plague of Athens in 430 BC, survivors attempted to provide help to the affected masses[44]. But identification of immune mechanisms had to wait for the dawn of modern biology. It was only in the late 19th century that linear antigen-antibody interactions were conceived. Shortly afterwards, Metchnikoff [1845–1916] disclosed the importance of phagocytosis. Through observation of a rose thorn penetrating a transparent starfish larva, he became convinced that the amoeboid cells attracted to the site of injury had evolved specifically to defend the starfish's integrity.

This chapter should allow the reader to understand the need for a specific cell system designed for defence. Sessile plants and animals, incapable of fleeing from predators, need onsite defence – mainly, primordial immune reactions and by releasing toxic substances of a diverse nature. It is on the basis of this observation that the evolution of immunity has to be seen; that is, mobile creatures retained primitive immune mechanisms in the first place, developing them into sophisticated systems.

CRUNCH!

FIGURE 12 PHAGOCYTOSIS EVIDENCED BY STARFISH
TRANSPARENCY PERPLEXES METCHNIKOFF

A landmark in the evolutionary process of immunology is coincident with the appearance of vertebrates. What is called the 'innate immune system' became supplemented by the 'adaptive immune system' that retains a specific memory for stimulatory determinants. A feature of 'innate recognition' is the ability to bind to the components that are common to different micro-organisms. The innate system, to which also belongs the complement's molecular cascade, always responds in the same way and time, irrespective of the number of previous exposures.

The ideas of immune recognition peaked by conceiving a foreign antigen dominated idiotypic-antiidiotypic network[45] (*see* Glossary). Step by step, cells differentiated into the different types, secreting, respectively, activating/suppressing molecules. Complex immune responses result in concert with an array of cells to which endothelial and even epithelial cells contribute. Links with organ systems are to be considered. Especially of note is when an inflammatory component becomes prevalent.

There is no doubt about an important role for immune reactions against pathogenic invaders; either manifesting or suppressing disease symptoms. The gap between both alternatives is most significant.

COGNITION-IMMUNE BIAS

FIGURE 13 PARIS' CHOICE OF BEAUTY WAS APHRODITE. WHAT ABOUT IMMUNITY: DO YOU HAVE THE RIGHT INSTINCTS, TOO?

Before starting, the following points may illustrate the aims of this section:

* until the silicon chip became available, carbon held the key to harnessing information (information meme);

* weak solutions may attract particles, while high concentrations repel (chemical recognition behaviour);

* long-standing habitat sharing may induce synchronization of menstrual rhythms (erasing differences between individuals);

* some species actually smell water, while others depend on true hygroreceptors (unlooked-for perception).

Regulatory mechanisms are in fact astonishing, especially when considering detection of deviation. Recent insights have now deepened our understanding of immune defence – but by no means comprehensively. The 'immune sense' adheres to what we call a 'complex order', while daily observation suggests that, like our brain, immunity strictly filters its choice of whether to focus on, or to ignore, a given signal. To induce the system to operate effectively, a large number of antigens usually have to be disregarded – while they may also present themselves in a camouflaged state. Only a few, significant in their context, will be selected for response. That repetition of stimuli leads to habituation is common in biology – its benefit has already been demonstrated at microbial level – thus, like persistence of heat-shock proteins (*see* Glossary), it has to be seen as evolutionarily advantageous. Because of the many variables involved, individual reactions are not exactly similar and may sometimes adapt poorly.

THE INSCRUTABLE INSTINCT OF SELF-PRESERVATION

> *Einstein did not rely on what Man perceives, so he reduced concepts like space and time to mere intuition.*

Habituation tends towards simplification and to automation, implying a learning process. Overall, however, it does not make much difference with instincts; except that the latter are inbred and may count on a long ancestry. The point is that automatic reactions are often astonishingly on target, reminding us of the saying '*tiene un Dios aparte*'. The outcome appears to be over-optimistic in statistical terms. Usually, the mental sphere seems to be involved in the first place, mainly when the subconscious is included; although, overall, reality seems to be better. Our idea is that more environmentally inclusive factors intervene. Besides atmospheric phenomena, cosmic radiation, etc., a holistic view has to account for a role by the immune function.

When we drive hours by car or in congested public transport, our senses – immune sense inclusive – keep an open eye. Most stimuli, be it a newly painted house or dangerous germs, remain unnoticed, many of them being stored subliminally. We only register something when the sense perceives it as a real problem, be it a serious car crash or dangerous invaders. In the former case it sticks consciously in our memory and in the latter it may trigger symptoms such as cold chills, shivering, fever, joint and muscle pains, even a sensation of near-death. In any case, it is the host who defines the level of alarm; on the other hand, the alarm may turn out to be a false one. The word 'inscrutable' is described in the dictionary as 'incapable of being investigated, analysed or scrutinized, impenetrable'; perhaps this 'Inscrutable' sense could have a role that is out of reach of our more traditional senses.

In immune defence, individuals' responses are often badly adapted to each other. Immunity is under the influence of many factors, often surpassing antigenic stimulation as such. Like conventional senses, the immune network has to become used to a routine. For example, we may not become totally aware of each of the details in our field of vision; nevertheless, many of them may remain stored. Subjects in a crowded habitat are steadily exposed to immune stimuli – often shared. From a biological point of view, individuals are connected in this respect. The question is, how? Routine exchange is unlikely to remain without consequences. One interesting aspect of the cognitive capability of immunity, in fact, is that, when responding, it is usually to a combination of different stimuli. Often, related immune-recognition chains lead to similar immune behaviour; and, therefore, to variants of the same response. For insiders, during their maturation process, it could have a role in immune response streamlining. In a shanty-town like milieu, this could be one of the reasons that lead to the observed common refractoriness against certain infectious diseases – but by no means all of them.

At present, it is not known if the hitherto overlooked 'noise' exerts any influence when an organism chooses to immune-react. Stochastic noise is ubiquitous and biochemical noise contributes to it. It is thought to relate to the highly complicated scenario of gene rearrangement in the lumen of the colon, preceding behavioural changes in bacteria – for example, to integrate a defensive colony or to invade the host[46]. Our interest points specially to eunoise, meaning the level to which biological functions become tuned. Stochastic resonance should not be omitted in any discussion about immune regulation and health homeostasis. It would be in line with nature's tendency to adapt to available network patterns[47]. There is no doubt that cognitivity enhances the ability to survive hostile environments.

Again, we need to remember that, similar to the usual lack of taking notice of customary stimuli, far more often than not the parallel not only extends to the removal of potentially dangerous invaders but to tolerance and even immunization against heterologous germs and antigenic molecules. Thus, rather than with simple consequences of exposure, here we deal with complex cognitive decisions as an integral part of the defence system inside the body.

It must be borne in mind that our IDS is under the reciprocal influence of other physiological networks. By (intelligently?) extracting information out of raw input and shaping experience, the capabilities of the immune system are shown to be very flexible; for example, 'tolerating' the implant of an allogeneic embryo means a major concession to the 'self–non-self' principle.

Immune subjectivity is not necessarily limited to the individual. Previously, we hypothesized that, by becoming critical, the mainly molecular inter-host traffic of microparticles is to be expected to repercute on the immunological autonomy. Our present view encompasses the community as a whole converting the subject even more complex.

It is worth mentioning that, besides being able to show tolerance (*see* Index), acquired immunity can also become misdirected – for example, by attacking 'self'. Similarly, reacting against harmless materials such as food, pollen or dust is as uncommon as it should be. It is obvious that, to be effective, not only is recognition necessary, but so also is an appropriate response.

Although much data has been collected about qualitative and quantitative particularities in understanding antigenic signals, it is still not always clear what drives the immune system to choose between straight dismissal of antigens, refractoriness, or to attack furiously. The rules of tolerance induction clearly require further research. It is becoming increasingly clear that in immune decision making, additional signals are involved which may not necessarily be directly related to the antigen. Perhaps individuals have the capability of sharing a common learning process.

The awareness of intercellular transfer of many molecules of the immune system, even in the absence of physical contact between the cells, in respect to our hypothesis, is indeed a step forward. Although a physiological role for this phenomenon has yet to be established, the exchange of molecules between cells may influence the immune system with respect to immune amplification as well as regulation and tolerance. In the case of the major histocompatibility complexes (MHC), Smyth *et al.*[48] suggest that the intercellular transfer of MHC and related molecules may have an influence on allorecognition and graft rejection; it should be remembered that skin and *mucosae* are no unsurmountable barriers either.

TOLERANCE

Accounting for species priority, in principle, intra-individual and group immune strategies should evolve in harmony. To suppose regulatory ruling is superfluous sounds counterintuitive; in the *scale invariant picture equilibria* tend to establish inside cell bodies and society.

Immune responsiveness diversely manifests itself and may evolve or vanish in phases. Immunologic tolerance regarding one or more specific antigenic determinants is manifested by diminished expression of either cell-mediated or humoral immunity[49]. Immune cognition is most convincing in oral tolerance, first detailed about one hundred years ago. Feeding a wide variety of non-pathogenic antigens can inhibit subsequent immune response. All systemic immune responses are susceptible, but the degree and scope of suppression depends on the nature and dose of the fed antigen. It may be the homeostatic mechanism that prevents hypersensitivity to food antigens. A similar process may prevent the aberrant immune responses to commensal bacteria that occur in inflammatory bowel disease. Precise knowledge to discern why one antigen may be ignored and others induce tolerance – in other words, the exact mechanisms responsible for the systemic tolerance and the reasons why tolerance is the default response to many antigens – remain controversial. Antigen selection may be of benefit, for example, if preventing auto-reactivity.

In early life, contact of self antigen with antigen receptors on immature B or T lymphocytes induces death or long-lasting unresponsiveness of these cells and thus clonal deletion[♥] or anergy. Such a state provides for further tolerance brought about by the absence of a secondary signal provided by helper T cells or by the antigen-presenting cell[50]. Conversely, it is likely that peripheral tolerance can result from multiple molecular interactions with antigens, although

---

[♥] Clonal deletion is the loss of lymphocytes of a particular specificity due to contact with either 'self' or an artificially introduced antigen. From clone (Greek Klon = young shoot or cutting), genetically identical cells or organisms derived by vegetative reproduction from a single parent.

diverse additional signals are often definitory. Tolerance does not mean that the role of the host is passive. The IDS maintains a reasonable control over whatever happens in the kaleidoscopic transit. Practically similar to the events inside the body, macrophages are endowed with receptors to preserved molecular structures that might suffice to identify and overcome bacterial cell walls or structures. Blood plasma provides proteins belonging to the innate arm; complement, C reactive protein (CRP), among others, also act against bacterial insult. In contrast to these pre-existing and evolutionarily very old structures, specific antibodies are produced through acquired immunity based on somatic gene rearrangement. Mounting an immune response against an appropriate antigen also means recruitment of accessory molecules that may act through adherence, or by delivering direct costimulatory signals. Many of them, we speculate, may be replaced by molecules from close subjects or even from other species, including microflora.

Allogeneic transplantation still poses questions, but is also a significant contributor in the understanding of the mechanisms of tolerance. Peter Terness recaps: one speculation says that once induced, immunological tolerance needs the permament presence of foreign cells in order to be maintained (see also microchimerism). If the allogeneic cells are removed tolerance disappears. In Medawar′s historical experiment the tolerant animals permanently carried the donor cells used for tolerance induction[51]. Because regulatory T cells are the main player in Medawar′s model, it is thought that the maintenance of their function requires the presence of antigenic stimulus. It seems that the form in which foreign cells are administered is pivotal for the induction of immune response or tolerance. If administered to an animal, foreign dendritic cells induce a strong immune response. However, if these cells are incubated *in vitro* before administration with certain chemotherapeutic drugs they become tolerogenic [52,53]. The chemotherapeutics down-regulate the expression of costimulatory molecules such as CD80, CD86 on the cell membrane and up-regulate suppressive molecules (e.g. ILT-3). One and the same allogeneic cells may act as an immunostimulator or immunosuppressant depending on the molecules expressed on its surface[54]. Apart from deliberately generated suppressive dendritic cells, it is known that the immune system harbours naturally suppressive dendritic cells[55]. Such cells can induce regulatory T cells which in turn inactivate the immune response.

As has already been said, cognition is central to immunity. Focusing on cognitive properties implies the connection of individuals with external factors. In a broad sense, externalism is more than the exchanging of messages; it might even serve to protect the weakest members from the group as a whole (the idea is also reflected by Figure 8). Connection allows us to acquire knowledge and to learn how to behave; it also favours fostering. We suspect a role for tuning immune defences. Besides that, most of the compounding molecules in bacteria relate to communication; there is now ample evidence about how micro-organisms collectively organize living – both defence and 'hunting'. From very low down the evolutionary scale[56], cognition starts with self-recognition.

The first question for a being is 'to react or not to react?', which is a basic principle in nature. In charge mainly are nerves (brain) in tandem with the IDS. Of course, signalling systems inside multi-cellular organisms with tissue specialization can be seen as more sophisticated – 'talk' among cells in the network context is very dynamic. Accounting for species priority, in principle, intra-individual and group immune strategies should evolve in harmony. As will be subject to discussion, information, by repetition, may be conveyed through cliché-like simplification and in the immune sphere this may be supported by packed conditions.

FIGURE 14 RIGHT AND WRONG

Modules

Like a cat will be recognized by every child, in the case of microbes a key antigen for response will be selected out of perhaps a myriad of so-called epitopes. Significant challenges to which we may become exposed may prompt the unfolding of cognitive and reactive 'modules' to speed up decision making and response tuning. A constellation with uncommitted B and T cells should not be considered as a *tabula rasa*; it may have a say in selecting suitable memes out of an intruder's complexity. We highlight this again in the belief that, besides cellular protagonism, other kind of elements ought to be considered, fitting in 'social immune' context. Here too, the virtual construct may be nourished either by a major impact or by stimulus repetition. Like an immune toolkit, as a module, may be ready not only on conventional innate and adaptive response[57].

Of necessity, immune cognition has to be severely filtered and the idea of cognitive toolkits seems appealing as well. These have been shown to be essential in diverse aspects of life. They may also be functional when opting to alarm or not. Unfortunately, alarm is also close to panic and self-harm. These aspects are especially considered when scrutinizing the 'pros and cons' of population density and sanitation as to what extent living comfort may in fact improve or modulate common health.

Immune Cognition in Body Belt Context

The milieu inside and around may change. Usually, an individual carries about ten micro-organisms per body cell. By assuming, paradigm-like, that everything in Nature interconnects; nowadays the '*lost environment*' seems so far to us 'behind the scenes'. The idea of an encounter—supplementary even functional – even a thinnest layer could have bridged supposedly providing response tuning assistance to the respective IDSs. Bearing in mind wide-ranging connections via pheromones and current acceptance of the apparently far-fetched Lorenz's butterfly effect, conventional threshold levels for immune interconnections should be reassessed (*see* 'Satellite Environments', page 99). It is on this basis that a possible minor, complementary, immune medium alongside the body is to be seen.

Interdependence Through Family Bonds

There is no need to go into detail about maternal ties, as they are shared by people in both the over- and under-populated areas of the world. The antibodies that the mother delivers, together with other immune-related molecules and mononuclear cells, efficiently contribute to the survival of the offspring. In association with natural auto-antibodies, the transferred support allows a relatively safe period for the child to evolve its own defences.

At birth, the active transfer across the placenta often associates with bidirectional pulses of blood. Such transfusions mostly are maternofoetal. Breastfeeding prolongs sheltering even more: among the many molecules, immunoglobulins are well absorbed by the versatile, relatively recently discovered FcRn receptors. Nursing implies 'support', denoting a wide-ranging immune interaction affecting the offspring.

Marital immune exchanges, principally through sexual intercourse, in a certain way also appear to be close to the envisaged community context – though there is no reason to believe that rural people refrain from promiscuous habits any more than do 'megacity' dwellers.

CHEMORECEPTION

A lot of the sensory processes to which organisms respond are induced by chemical stimuli as a main factor[58]. For example, after an initial interaction with a cell receptor, before becoming converted to nerve impulses, the incoming stimuli are channelled through nearby cells, referred to as secondary receptors. By interacting with key cells, chemical stimuli provide much of the information about the environment. While it has been accepted for a long time that microbial invasion not only depends on the pathogenic condition *per se*, location, intensity, duration, etc., have also to be taken into account. Although correlated, less attention had been paid to interhost chemical connections in the microhabitat. Much has yet to be done to decipher physiological mechanisms of chemoreception and their role in patterns of accommodation to a hostile environment. Materials, especially gases, may come a long way before reacting with so-called distant chemoreceptors, whose thresholds are usually very much lower than those of contact chemoreceptors. Among the behavioural results from distant chemoreception, biologists assign importance to orientation in a wide-ranging sense.

It has been held that invertebrate animals have only one chemical sense, with different sensitivities for various chemicals; whereas terrestrial invertebrates such as insects exhibit separable chemoreceptive capacities. Aquatic animals and terrestrial species with mucus-secreting skin are generally sensitive to chemicals all over the body. This sensitivity has been called the 'common chemical sense'.

Often chemical receptors are usually free nerve endings (branching structures or dendrites) and common chemical sense is different from pain induction and thus is considered as a separate sensory capacity. Man and other terrestrial vertebrates have a remnant of this receptor system that responds to irritants in the mucous membranes of the mouth, eyes and genital organs. Although chemoreceptive faculties among birds with developed taste buds and olfactory brain centres have hitherto been assumed to be deficient and subordinated to the organs of visual and aural perception, that consensus is now being reassessed.

CONTRASTING MEANS TO COMMUNICATE

Why common grounds seem hard to recognize

Two days after the tsunami hit, he took out his bow and shot an arrow towards the rescue chopper – a signal the Sentinelese from the Indian archipelago of Andaman and Nicobar have sent out for millennia: tribesmen want to be left alone. 'To survive, they needed to learn nature's sights, sounds and smells – even gauge the depth of the sea with the sound of their oars. A sixth sense which we don't possess.'

Neeles Misra
Ass. Press, Dec. 26, 2004

Sight and hearing confer an enormous advantage to the nerve system in external connections. There is no doubt that it obscures the body of otherwise most interesting equivalent functions shared by the IDS and the nerve system. Such cloudiness is not surprising when considering the priority of group cohesion. Possible external immune crosstalk is supposed to depend on molecular and cellular interactions – released from the repertoire of the respective host-IDS. Mainly, it is to remind the *'milieu soup'* (see milieu-soup-interface) that may evolve as body-wrapping film. This may have been the case of our forebears, provided that they inhabited crowded population centres. In the shanty towns, as their historical remains, residents often continue to be subjected to almost equivalent conditions. In this context, the *'lost environment'* may have a special meaning.

FIGURE **15 H**ERE SIGHT UNDERSCORES INTERHUMAN IMMUNE COMMUNICATION HANDICAP AT MOLECULAR LEVEL; BRIDGING BY A CRITICAL MILIEU SOUP MIGHT BE A POSSIBILITY

TECHNOLOGY COMES ON STAGE

. . . rapidly attaining the status of paradigm; right now, we witness the vogue triggered by the cell-phone mania, taking away substantial sums from the already scarce income of underprivileged people. The present fanaticism is only one of the examples revealing the magnitude of the role that technology plays in our life. Continuous renewal is another feature that characterizes the current wave. Uncertainties that the proviso implies notwithstanding, we share the current enthusiasm. No doubt that technology contributes to improved quality of life.

It must be underlined that the contrasts quoted in the means of inter-human communications can only be seen as relative. A third level of connections, namely, the many faces of informatics, has already become engaged in zooming external body connections. Also, because of TV, the internet, etc., the stakes are high. It is possible that, sooner rather than later, the whole scene is

69

likely to be under technological control. It doesn't sound strange any more that we will see and even talk with astronauts who are expected to visit Mars in brief. Here too we think that insight about light and sound waves by connecting brains precedes an immunological parallel outside the body.

PHEROMONES

The complex story of biological means of communication has, in pheromones, a hallmark exploited by many species. The idea of the existence of human pheromones (*see* Glossary) dates back to the 1700s, although for a long time it was denied because no proven evidence for association could be found; the identification of suitable receptors being rather recent. Already the etymological root signals the difference with hormones; by being launched externally they may carry out their respective specific task at most diverse distances. These hormone-like substances are remarkable for their diversity; their multiple functions in particular. Animals, either in groups or dispersed, depend on their cognitive ability. Pheromones have long been associated with olfactory perception; for example, honey bees scent mark their own hive and areas around it with odours that restrict identification to the members of their particular community. By acting as important regulators of social behaviour, some pheromones may be crucial in identifying insiders, excluding others. Besides territorial instincts, pairing, menstrual synchrony, etc., also belong to the aspects of life that species-specific pheromones may affect. Alarm odours alert ants, wasps or bees when colonies are endangered – in a way similar to toxic or repulsive chemicals that plants may emanate for defence. But pheromones are not only revealed to influence a precise behaviour in the animals of the same species; the chemical communication can also involve inter-specific, even inter-kingdom (e.g. between plants and insects) communication.

In biology, a conditioned stimulus response cycle is rather common. In the search for immune communication analogies, it should be noted that pheromone-binding proteins appear to participate in chemosensory coding (cross-reactivity was not related to taxonomic relatedness of the species, but rather to chemical relatedness of the pheromones to which receptors from these species were tuned[59]).

Previously to receptor activation, many pheromone-binding proteins (PBP) appear to have a filter function[60]. Usually, protein kinases become activated upon pheromone stimulation; however, some variants still require elucidation. Pheromones have been shown to activate gene expression in cells from many kinds of animals, humans included, thereby inducing the release of the respective functional substances. Among others, stress conditions or apoptotic agents become involved early on in pheromone-responsive signal transduction cascades[61].

Pheromones have been shown to be active at extremely low concentrations – often threshold levels in oceanic ecosystems are very low, prolonging the distance over which encounter dynamics take place[62]. Such is the case of Schreckstoff (German: 'fright substance') that is given off by agitated fish[63]. The fact that the chemical signal is able to alarm others of the species at a radius in the order of miles tells us that we are dealing with solidarity in defence that is widespread in nature. To link such solidarity to a putative collective immune support system appears to be entirely appropriate. The response to sudden infectious challenges may vary between individuals, but this does not deny the possibility of mutual influences between them. It might be thought possible that, for the individual, the acquisition of information about invaders at an early stage may improve the capacity to counteract their virulence.

As a supposed attraction for the opposite sex, pheromones have now clearly shown their position regarding the cosmetic industry – but, accounting for science in expansion, the turn for collective immune influences' research in depth may also not be too remote.

RESEMBLANCE TO UNICELLULAR BEINGS

Firstly, a brief comment about the particular, evolutionary-grounded relationship between microbes and us, humans. Since time immemorial, the cells from our organism conserve molecules that are similar or almost similar to those integrating germs. Interestingly, we use them as a means of defence against them. In certain fundamentals of life, microbes are streets ahead of, for example, vertebrates. These minuscule beings have been shown to be far more plastic in finding ways to attack and to defend. This depends mainly on their amount in tandem with extremely brief inter-mitotic interval that definitely favours the incidence of mutations, thereby conferring substantial evolutionary benefit[64]. However, quite often host defence mechanisms tend to keep pace – be it by somatic mutation, phenotypic plasticity or, last but not least, commensal microflora support. But the speed in acquiring drug resistance is to be seen as a most considerable problem. Microbial versatility has more under its sleeve: for example, chameleon-like, antigenic determinants disguising.

Microbes form colonies more often than previously thought and do take advantage of this situation; the apparent freedom by which countless creatures may disband is not to be underestimated either. Even the very simple ones that apparently lead an independent existence, when sensing danger show a strong tendency to form colonies. Switching to become part of a colony not only implies the reshaping of their defence mode. In the compact masses, often difficult to distinguish from simple structure-lacking multicellular beings, conduits may become outlined for nutrients or waste; importantly, communication mechanisms may enable them to behave as communities. In effect, many of the molecules that are secreted by microbes appear to be able to communicate with other organisms – not only of their own species, but also with those of other species, even with far more complex organisms. The spec-

73

trum of information exchange ranges from rudimentary defence to symbiotic associations, some of them surprising in their sophistication. Population density plays a key role – 'quorum sensing' (QS)[65] signalling allows to define gene expression[66]. For certain microbes in close proximity, QS is a most valuable tool. It comprises proteins that show distinct responses in terms of the timing or level of protein accumulation, or the direction of change. They may also downregulate genes, allowing microbial populations to evade the IDS, maintaining a low profile until sensing sufficient population to start up an attack. Conversely, in response to real or perhaps imitated QS signals, hosts have shown to develop ways to invalidate QS molecules.

We are now paying more attention to the benefits of microflora and even to regulatory influences from livestock with which man has more recently been associated in his evolution.

Recent evidence suggests that microbes are able to build ecosystems even more sophisticated than that of the Amazon. Where? Just inside our bowel! The question arises, to whom in fact does it belong? – maybe we represent to them no more than a mere shell.

Limits For Association

That the environment contains many potentially destructive micro-organisms is obvious from the speed with which animals decay after death, although many microbes live harmlessly in contact with their hosts. Normal microflora is abundant on skin tending to concentrate at orifices – the important point is that the inside of the body remains sterile (although there are exceptions[67]), and the immune system acts to this end against invaders whether breathed in, penetrate the skin by punctures or cuts, or gain access by other means.

The relationship established by pathogens and their target is not exclusively one of predatory harm. It may affect the whole immune equilibrium of the host, with the involvement of other networks in the body. Possible repercussions outside skin deserve to be carefully explored .

It seems likely that immune exchanges between the individual and his surroundings are a constant event, although for practical purposes the effective radius of action appears to be extremely restricted – the reason why it tends to remain mostly unnoticed. Nevertheless, over time the milieu might exert some influence on the immunological autonomy from nearby members of the community and if so, this of course would not totally preclude a more expanded area either. Bearing in mind the importance of ecological links in matters concerning immunology, it may be a two-way street. Other questions are: can the external human environment contribute to the image of 'self'? Is it biologically acceptable that identification and defence exclusively centre in 'self'? As yet there is no conclusive proof that immune-subjectivity exclusively belongs to the individual.

ANTIBIOTICS AND PESTICIDES

FIGURE 16 ANY RESEMBLANCE TO THE FATE OF ANTIBIOTICS
IS A MERE COINCIDENCE

When in the 1950s antibiotics became popular, optimists uncorked celebratory

champagne as 'masters of nature'. But what is thought of as fundamental can change. The increase in antibiotic-resistant infections and the shortage of new breakthroughs in the field have inspired our cartoon showing the expectations of retired people versus the ability of the shrinking workforce to afford to maintain them in spite of an ever-longer lifespan associated with skyrocket health-care costs. The plethora of antibiotics, at least for now, is failing. For example, *salmonella* exhibits a high rate of plasmid-related antibiotic resistance. Often a transposon encodes a β-lactamase that is inserted into either a plasmid or a bacterial chromosome. These self-replicating circles – containing DNA – called plasmids, are best known medically for their ability to confer antibiotic resistance on host bacterial strains and to transfer this resistance from one bacterium to another. Multi-drug resistance may result and has already posed serious problems, even in hospitals in the developed world[68].

Quattordio (Figure 17) reminds us that, even today, indiscriminate prescription of antibiotics lingers and established interest groups cannot easily be dismantled. The same can be said about careless administration of pesticides and fertilizers by a significant proportion of veterinarians and farmers (see further on). Fortunately, by expanding insight into molecular biology, friendlier solutions are to be expected.

The speed by which drug resistance evolves signals microbial success. Because of their immense numbers and dispersal, microbes convert many triumphs of the antibiotic era into a Pyrrhic victory – proving that disease cannot be eradicated, as had been thought. Microbes that manage to survive the effect of a promising new antibiotic have been shown to shed mutated molecules which are incorporated by others of the same or different species – implying that nosocomial protocols are at risk. So far, no overall agreement that the antibiotic era may be counterproductive seems to be justified. Even when treatment proves to be only partially lethal to the microbes, this could possibly come in support of the IDS. Some treatments are running on this premise; for example, for HIV-infected patients. Nobody believes that antibiotics are immune-stimulators; on the other hand, any debilitating action upon microorganisms gives way to a proliferation of the survivors, in particular those that had mutated to resist the antibiotic.

FIGURE 17   ANTIBIOTICS AND PESTICIDES EUPHORIA IN RETROSPECT: MORE HARM THAN GOOD?

We focus again on insecticides: even if harmless to people, they may bounce back, for example, by destroying predator insects that in Nature would keep pests in check. This handicap may be coped with by rearing predators that have gained resistance to the pesticide – such strains could be useful in the attempt to re-establish the ecological equilibrium. It follows that, tit for tat, an unwanted outcome sharpens resourcefulness.

Since time immemorial immune systems have developed a well-honed perception, not only for invaders but for manifold signals as well. As far as is possible, immunity's reliance on cognitive properties should be emulated by therapeutics. Detailed knowledge of the ways in which the immune response priority selection takes place is important. The accepted ideal is to attempt eliminating danger without affecting anything else, thereby acting in a regulatory capacity.

As acquired refractoriness to antibiotics and drugs are more than mere hints, there are many problems to be solved. On the other hand, suitable treatments for new threats may emerge. We must uncover all possible means by which microbes disseminate multi-drug resistance. Furthermore, complete information of predatory activities and how invaders coordinate and multiply are needed. Here we remind the reader that we see the glass being 'half-full' instead of 'half-empty'. Public health interests already tend to be served with more sense of responsibility worldwide. But, as the Bible puts it, prophets were often lonely.

DECOY – CAMOUFLAGE

Camouflage is deeply rooted in biology. Many animals, plants and micro-organisms use camouflage to attack or to defend. And from another perspective, camouflage is often a role for the host. Antigens can be identified either nude or clothed.

FIGURE 18 A ROLE FOR THE COMPONENTS FROM THE HOST'S FLORA MIGHT BE TO DIVERT THE PATHOGENS FROM THEIR ORIGINAL TARGET

At times, certain largely innocuous germs may react in an immunomodulatory fashion, possibly tuning into a connection with the commensal flora. This means that, in addition to nude hetero-antigens in the *milieu*-repertoire, our IDS has access to shapes of jointly connected or perhaps integrated molecules that may act as supplements for response induction. Persistent re-circulating antigens and minute inter-individual antibody-antibody interactions could count in fluent exchanges of immune-cognitive stimuli. Interest in decoy functions in the body is on the rise. Nydegger includes antibody catching soluble Fc receptors into the idea of a 'first line of immune defence'[69]. By competing with their cell-bound counterparts they may prevent the phagocytosis of antibody-coated platelets. The common cold remains a heavy burden in terms of absenteeism and finance – proper treatment is still not available. Among the attempts to interfere with rhinovirus access to cells covering the upper respiratory tract we may cite decoy strategies such as ICAM-1♥ molecules-based vaccines. Modern warfare technology profits from similar principles; for example, fighter planes, by scattering aluminium particles, confuse radar systems to deflect anti-aircraft fire.

In the light of these developments it seems attractive to consider that the IDS is probably somehow related to decoy functions by the molecular and microbial repertoire around the body.

♥ Intercellular Adhesion Molecule.

I - TRAPPED

> *"If history of mankind demonstrates anything at all it is the extremely slow pace taken by the academic and critical mind in recognizing the existence of facts presenting in disorder, without pigeon-holes or compartments, or events that threaten to break the accepted system."*
>
> William James, at the dawn of the 20th century

The subject deals with disparate topics, 'discrimination' being the common denominator. The section is set apart because – in context with the IDS – discrimination, as a suggested 'biological property' in the behaviour of most species, often fails to contribute to protection.

MICROCHIMERISM

*Do promiscuity and filthy
crowding belong among its
potential causes?*

With the advent of organ transplantation, co-existence (chimeras) came to the fore. More recently, small populations of cells or DNA derived from a genetically distinct individual have been found; but, since Dausset, the venerable precursor of molecular specificity of individuals, suggested that HLA-specificity is to prevent microchimerism of any kind; it took time before it was taken for granted that long-term microchimerism is rather common. The appreciation of naturally acquired allogeneic cells is relatively new. In any case, after 'contamination', foreign cells may persist for decades. HLA-specificity can be circumvented by tolerance; for example, foetal immune immaturity. While co-existence after bone-marrow transplantation usually requires downplaying of both host *versus* graft and graft versus host immune reactions by therapeutic means, at a minor scale spontaneous transfer of allogeneic cells has also been identified, either from the mother to the foetus (foetal microchimerism) or from the foetus to the mother (maternal microchimerism). The immature foetus is of course more tolerant although, occasionally, long-term persistence of foetal cells in the mother may have significant repercussions. Twin-twin transfer in the uterus is another possibility. Iatrogenic microchimerism after blood transfusions has been repeatedly reported. The lack of tissue compatibility remaining a major obstacle, it is surprising how often minimal amounts of surviving allogeneic cells have been demonstrated[70]. Considering organ transplantation rejection propensity; how does the rather frequent microchimerism escape immune *surveillance?* (see Tolerance).

Long-standing transfusion-induced microchimerism is in fact a subject that deserves special attention. For practical purposes, too many factors require a careful evaluation and standards of transfusion intolerance remain far from consistent in different parts of the world. It is still almost impossible to verify if transfusion symptoms simply derive from pyrogens (fever-producing contaminants) or if the cause is a specific sensitization against blood cells (any kind) or against a plasma components. Furthermore, in population clusters, the heterogeneity of MHC molecules differs dramatically. Finally, conditions such as climate, nutrition and others have to be taken into detailed consideration as well.

Mainly on the basis of an unhygienic crowded milieu*, we think that jumping from host to host may be another non-iatrogenic source of microchimerism. By multiple body contacts it seems likely that some cells become transferred via secretions, excrements, etc. The problem is that the shantytown subject does not seem to attract such a line of investigation investment.

Microchimerism has been repeatedly related to autoimmunity; even a cause-effect has been postulated[71]. But, as the incidence of autoimmune disease after organ transplantation does not tend to increase substantially, how can we explain the discrepancy? Prevention of microchimerism is not on the agenda so far; the scope will broaden, if links with 'so-called' autoimmune disorders become realized.

The traffic load is thought to hold functional immune molecules and a number of cells or suitable genetic material, carried perhaps by apoproteins or bacteria. Penetration of the heterogeneous complex may be facilitated by lesions, although epithelial and mucosal cell linings cannot entirely stop it either. We may believe that after the odysseys some cellular material will remain viable. If so, babies and immunodeficient subjects are to be considered in the first place. As constrained bed-sharing preferentially affects siblings and parents, genetic links – often a coincident haplotype – could play a role, maybe compensating for the handicap resulting through lacking a venous

---

* See more in 'Pollution – Milieu soup-interface', page 102.

access. So far there is no proof that this may lead to long-lasting microchimerism.

FIGURE **19** **G**UESS WHICH ARE THE ELEMENTS FROM THE COLLEAGUES WHICH MAY CONTRIBUTE TO OUR IMMUNE NETWORK

CORNERING MICROBES

In spite of the ever-present food chain, many microbes behave neutrally or may even display mutualistic features; but, mainly in genetically susceptible hosts, they are capable of inducing disease – chronic and so far unperceived inclusive. As viruses are dependent on host survival; also on dissemination systems in which vectors play a role, suspicion to link chronic disease with viral strategy to perpetuate as revealed in the literature cannot easily be dismissed.

There is always a probability that micro-organisms tend to attack in huge numbers and, indeed, don't care about casualties. As has been discussed, freely dispersed microbes may also change to become part of a colony, thereby reshaping into a defensive mode. As a third possibility, out of reach from being decimated by the IDS, microbial ADN detected by PCR suggest certain common types may survive in 'reservoirs'[72, 73]. *À la long*, the putative 'forced' confinement could link to progression of arterial plaques. Such microbes, once perceiving a weakened IDS, may not miss the opportunity to take up an aggressive mood.

AUTOIMMUNITY:
A PRIVATE CONDITION?

Healthy individuals have a 'stock' of auto-reactive antibodies and T cells which mainly originated early in ontogeny. These antibodies are already present at birth, reactivity being characteristically poly-specific with high connectivity. Such development takes place under the influence of the external environment, growth and functional maturation of secondary lymphatic tissues. Physiologically, they are supposed to be involved in immune homeostasis and self-tolerance regulation; but nowadays this kind of equilibrium is under challenge. Success and failure, the two sides of the coin, have only recently revealed a bias towards autoimmune disease.

Autoimmune disease *per se* tends to be considered as an immune deficiency, often as a sort of by-product of an immune-deficient standing. Manifestations may arise when clones are inadequately connected to the network. Thus, not all immune-deficient individuals develop autoimmune manifestations, nor can we be assured that every affected subject has an immune handicap[74]. Although there is ample evidence for a genetic role in autoimmune disease, only 34 per cent of monozygotic siblings from patients with rheumatoid arthritis [RA] will become affected by the disease and in no autoimmune illness does the percentage surpass 70 per cent. This might indicate that, in many cases at least, environment has a part in the development of the disease[75].

Still, the transition to pathological autoimmunity is often misleading. We should remember that, generally speaking, target antigens are dominant.

To evaluate clinical evolution, rather than particular auto-aggressive antibodies or cells, a complete answer may provide common immune features, provided they become properly identifiable. An appealing question is why clinical immune manifestations tend to show preference for certain body parts. For example, RA affects certain joints, while leaving others alone. Besides, variations in inflammatory intensity, duration and migratory propensity round up. RA is not always limited to the joints, although more often extra-articular manifestations are seen in patients with systemic lupus erythematosus [SLE]. A possible answer to the puzzle could be to link autoimmune symptom spots to the immediate multifactor milieu, and even beyond skin. Unpredictability persisting, practically no organ system or function is safe from being struck. The apparently capricious flare-ups and downs that often characterize the symptomatic course point to changes in the individual's immune behaviour. Stress may exert an influence[76]. As mentioned, a certain infectious agent or perhaps facultative pathogen habitué to the commensal microflora may be relevant. It has been postulated that infection is part of the initiation of autoimmune disease with no active role in clinical manifestations later on. To complicate matters, a different hetero-antigen or super-antigen (*see* Glossary) may trigger them. On the other hand, a significant number of reports inversely relate autoimmune diseases – mainly atopy, asthma, also RA and others – to infectious events[77]. Accordingly, antimicrobials do not belong to the conventional autoimmune drug armamentarium – although antibiotics have been shown to be efficient in certain clinical variants such as chronic inflammatory bowel disease; they may prevent exacerbation by inhibiting bacterial tissue invasion and decreasing concentrations of detrimental enteric bacterial species. The association with probiotics – commensal bacterial species with beneficial physiologic or therapeutic activities – may be useful to maintain the disease quiescent. Could microbial competition have a role? So far, supposedly because of genetic and environmental involvement, the answers to all the questions remain open.

Dynamics – Internal – Worldwide

In many cases it has become clear that autoimmune disease does not depend exclusively on a patient's state as such; to think about a direct environmental effect is almost certainly an oversimplification.

For decades – since the eminent Paul Ehrlich forwarded the '*horror auto-toxicus*' paradigm – the real dimension of the autoimmune problem had not been duly taken into account. Without any doubt, autoimmune expression affects ever more people, while historical chronicles rather refer to an affluent minority. Concordant perhaps with the 'hygiene hypothesis', Japan and Sweden belong to the countries with the highest incidence of both allergic and regular autoimmune manifestations.

To attribute the 'autoimmune wave' exclusively to industrial pollutants is risky, to say the least. Interestingly, in the final days of the former Soviet Union, despite being more exposed than their Swedish counterparts, the Estonians' rate was, in fact, lower. The confusing ethnic mixture could not account for a genetic explanation either. Concerning pathophysiology, clinical significance, and treatment of autoimmune phenomena in AIDS, many questions remain open.

Referring to the heading, autoimmune aggression can acquire characteristics that suggest an internal storm. Noël Rose estimates that, in concert with cardiovascular ailments and cancer, autoimmune disease belongs on the list of main human scourges. For an understanding of current changes in health, a look at broader foundations is required. Long-term non-progression in HIV-1-infected patients may belong to them[78].

AIDS Tragedy

"Long-term non-progressors are telling us we're missing something big with regard to how the immune system works"

Mark Connors[79]

The immune handicap in advanced AIDS often associates with autoimmunity – less frequent during the earlier stages of the disease[80]. The observation of HIV-associated (auto) immune phenomena dates back practically to the moment the disease as such had been identified.

Multiple factors may be implicated in derangement of the immune system's homeostasis, leading to exacerbated function and expression of pre-existing auto-reactive elements, despite the fact that, from a clinical point of view, the patient simultaneously has an immune suppression. Discussion still lingers as to whether current antiviral treatments modify DAT+[♥] occurrence in HIV-infected patients[81]. Anti-erythrocyte antibody detection increases with disease development and also rises the more patients are heaped together, so that physical contact becomes unavoidable (here we omit consideration of sexual conduct). Independently of the immune status, a distinction between forced confinements in, for example, a prison cell and optional confinement as, for example, a shanty-town dweller is at hand.

Among the life-endangering challenges as they happen in the Francisco Javier Muñiz Hospital (Buenos Aires), AIDS patients are subjected to different infectious agents, poly-medication, etc. They show a clear tendency to develop auto-antibodies and, mainly in the environment in which terminally ill AIDS patients are being treated[82,83], manifestations become prominent. The hospital for infectious diseases covers an area of 8 hectares that includes a

---

♥ Direct Antiglobulin Test to identify antibodies bound to red cells.

prison building for convicted criminals with AIDS. Considering that treatments were rather chaotic – be it due to lack of patient co-operation, irregular drug supply (not necessarily out of stock) and follow-up – the many and most severe complications made things much worse.

Although the hospital was overwhelmed by around 150,000 admissions a year and was in very bad shape, Dr. Carlos A. González and his team, in response to my suggestion (1997), hastily collected data of positive immune anti-erythrocyte antibodies (DAT+) in the three groups of AIDS patients, all of them belonging to stage IV-A to IV-D (classification proposed by the CDC[84]). This provided a possibility of comparing outpatients, hospitalized patients and prison inmates at similar stages of the disease.

Interestingly, González observed highly significant differences in DAT positivity frequency among the patient groups. Prison inmates showed the highest incidence and outpatients the lowest, while the data from regular hospitalized AIDS patients were practically equidistant in the middle (personal communication). The as yet unpublished data show convincing evidence of a three-stage DAT+ incidence in accordance with the degree of crowding. Although recent statistics fail to relate DAT+ incidence in HIV positive patients with haemolysis[85], considering the impact of the worldwide scourge, the observation of González and co-workers deserves to be taken into account.

II - TO CONSIDER

FIGURE 20 ARBORVITAE: FOUR EQUIVALENT FACTORS
REVEAL THAT NATURE TENDS TO SIMPLICITY

IMMUNE PARADIGMS

GEO-CENTRIC   HOMO-CENTRIC   IMMUNO-CENTRIC   AREA-CENTRIC

FIGURE 21   SELF-CENTRED PERCEPTION OUTMODED?

Figure 21 depicts the human-centred view as a natural tendency; in our case, specifically pointing to immunity. Dogs mark their territory through particular pheromones that they spray when urinating (see Pheromones). Chemistry, therefore, appears to be involved, but this is merely the starting point for further studies. There is more – cognitive scientist Douglas Hofstädter has commented on a computer program that performs Mozart's '42nd Symphony' (Mozart only wrote forty-one): 'Technique has no model whatsoever of life experience, has no sense of itself, has never heard a note of music, has no trace in it of where I think music comes from. I am comparing that with an entire human soul, as forged by the struggles and travails of life, and all the experiences that create emotion'[86].

When the establishment switches to a new paradigm, it often happens that novel concepts are given undue importance and it usually takes some time for a proper balance to be resumed. In this regard there seems to remain enough space for more or less fashionable ideas, 'self' among them.

After the emergence of the bourgeoisie in the course of the 19th century, typically, Henry Ford* can be seen as a symbol of the rising hierarchy of the identity of the individual; it became further emphasized in the 1920s by psychology. Gradually, post-Freudian* thoughts evolved to a much broader approach that attributes a major role to community. Likewise, it may be felt that, by swinging the pendulum back at the end of the 20th century, the immunological homunculus suggests a rather similar overestimation.

In the succeeding pages, immunology adheres to what we call a 'complex order'. In the context, more or less fashionable ideas – 'self' among them – deserve to be considered. 'Self' is thought to be monitored through a compound network of natural auto-antibodies interacting with an increasing variety of factors produced by cells that, by being stimulated, may release cytokine combinations that defy paradigm. Actually, the host selects a particular major auto-antigen as characteristic for a certain organ and tissue. Also, regulatory mechanisms and the astonishing ability to detect deviations are often attributed to 'self' – although exhaustive deciphering still requires further research.

We must consider the individual as a particular aspect of community, like time and presence, which were paramount in ancient Greek thinking and tradition. Looking at the individual in the context of community closes the hermeneutic circuit. It also follows that, in immunology, a particular case has to be referred to the whole[87]. Gene-dependent characteristics point to autonomous determinism independent of external factors – but only at first sight. Already,

---

* [1863–1947].
* Sigmund Freud [1856–1939].

inside the individual, a co-ordinated immune function also depends on impressive numbers of environmental variables tending to conceal an identification of common immune effects. Our aim is to look after such impacts through the surrounding environment. Conceivably, it could happen by direct means or via gene rearrangements; the latter do not necessarily have to be opposed to built-in genetic mechanisms.

As the IDS is subjected to many variables, individual reactions cannot be expected to be exactly identical; sometimes, it is of no surprise when they markedly diverge – occasionally causing them to adapt badly.

In the community context, possible mechanisms of collective defence regulation remain to be defined. They are now being approached from distinct angles. The notion of 'herd immunity' is also to be considered as a group phenomenon[88]. For example, in a group, vaccinated people can act as a sort of 'firebreak'. Much appears to depend on the extent of exposure, the 'herd' as such and disease-spread idiosyncrasy.

AN ATTEMPT TO REDEFINE THE IMMUNE SPHERE

FIGURE 22 EBB AND FLOW: DOES THE MOON'S EGO TOUCH OUR SEAS?

Irun Cohen had good reason to assume that the immunological homunculus mainly expresses some of the host's long-standing antigens[89]. By denying that anatomical body limits are absolute, the possibility arises that substances from the proximity may play in the virtual 'immune ego'. We think mainly about flora components: those normally present or those resulting from convergence. If true, the immunological homunculus would not be symbolized any more exclusively by selected host-specific molecules. The 'homunculus' notion was introduced by neurologists to outline certain neurons normally related to 'self'. The facts that the notion of self is relative and that it also depends on external factors were exploited by the Nazis, as can be seen from the damage they inflicted on their victims' egos by ordering them to shed their clothes.

Lessons from defensive co-operation among micro-organisms and interspecies forest dynamics also led us to question an immune-centric view.

COMMUNITY INTERACTIONS

Biological beings tend to be in touch

FIGURE 23 DOES THE MILIEU IN CONFINEMENT INFLUENCE IMMUNE BEHAVIOUR?

SATELLITE ENVIRONMENTS

The belief that only the body content calls the shots with regards to immunity is widespread and, besides inside the gastrointestinal tract, it is common to disregard the details of the space that surrounds our body limits. Here it seems appropriate to look at human biodiversity and how our immune structure entangles with the immediate outer milieu and even further. Among the raisons d'être to consider the possibility of an additional immune medium alongside the conventional tissues and body fluids must be current acceptance of the apparently far-fetched butterfly effect, the possibility of distant links via pheromones and possible other means as well, ubiquitous stochastic noise among them. The problem of interactions and the resultant complexity is related to an unexpected phenomenon, the non-decoupling of very different length-scales in some biological situations. Here should apply the notion of universality in physical systems with a large number of degrees of freedom. The question is that conventional information exchange thresholds have to be reassessed.

The term 'satellite' in this context has a special meaning. Think about a dual role for environments we carry (or have carried) with us. Beyond epithelial linings, a dense milieu soup (see later) ought perhaps to facilitate certain functional relationships in the close neighbourhood. Why not imagine that a feeble heterogeneous formation could itself also be of some influence on the supplying host? Think about human-made orbiting satellites that not only explore the universe but also focus on the earth for weather, spying or other missions. If so, shanty-town dwellers devoid of safe water and hygiene could count perhaps on an additional line of defence – pathogens may be either directly confronted at body surface or they could be attacked from the back.

INSIDE THE DIGESTIVE TRACT

Already, Louis Pasteur* and Ilya Metchnikoff (*see* 'Early Evolution of Host Defences', page 52); the legendary pioneers of immunology, attributed to our microbiota a significant regulatory role in maintenance and recovery of health. 'Microbial tolerance' has evolved in primates in concert with their intestinal commensal microflora, a situation of mutualism that has begun only recently to be unravelled. In the gut the archaic property of tolerance is paramount not only for food but to discern microbes of many kinds. Both innate and adaptive immunity provide the intestinal barrier with a combination of sensing and defence alert that permanently tends to counter intrusion of commensal micro-organisms, thereby generating a situation of 'physiological inflammation' Accordingly, not only pathogenic but also commensal micro-organisms can invade and disseminate in the host.

The traditional microflora has an important role in the maturation of the gut and its immune function. Still, to interpret individual variation within what is considered to be a normal range is difficult. Because of its magnitude, the microflora as a metabolic factor cannot be taken lightly. By its genetic diversity, the gut microbiome may surpass our genome a hundredfold[41], enabling it to perform scores of functions complementary to those from us – essential for survival for some of them. For example, it has been shown that the flora in the gut provides certain protection against colonic cancer. In spite of flora's resilience, the sparing of antibiotic treatment may not be meaningless.

The biomass settled in the colon – 700 to 2,000 gram – contains only 500 to 1,000 selected strains. However, due to their wide variety of responses, they provide enough diversity and redundancy to ensure customary robustness. Being a contributing factor to the host's stability, it is also involved in local functions such as transit time, bowel habit inclusive. Aiding the digestion of complex dietary sugars is even more important – a consequence of co-

---

* [1822–1895].

evolution of animals and their microbial partners. There is more: like tumour cells have shown angiogenic potential, the commensal flora may induce neovascularization to optimize nutrient absorption[90].

The host as counterpart gets in touch by the surrounding cell linings supported by abundant lymph nodes full of immune cells and their secreted products. The process of 'oral tolerance'[91] is sustained by a complex integrated structure to ensure an effective admission of nutrients on one hand and, on the other, defence against ingested toxins and pathogenic bacteria. Dendritic and intestinal epithelial cells are the main protagonists in the 'crosstalk' and interactions with luminal bacteria[92]. There is ample consensus about synergistic co-operation between the innate and the adaptive arms of immunity. It is also known that the liver, gall bladder and pancreas exert influence – although a quantitative judgement is still very difficult to make.

Coming back to the microbial ecosystem in the bowel, progress is being made to understand the many coupled interactions and feedback loops, and the evolution of living components and resulting change at the system level (*see* 'Complexity Theory', page 10). When a strain turns aberrant, it tends to be down-regulated by the community *ad hoc*, supported by immune component elements from the host. If this fails, the solution may come with a subsequent generation. At birth, the new acts of colonization depart from a small founding population derived from the mother – although the new round of selection will again be ruled by the host, once its pertinent resources have matured, in tandem with the newly evolved microflora. Obviously, the respective genomes are involved.

The interplay with the mass of the intestinal flora gives evidence that our bowel cannot be seen as a collection of isolated objects. With regard to the milieu of the gut, the links between its content with the epithelial linings represent a complex multifaceted encounter that provides us an idea of the interdependence to which the enteric flora and the IDS are subjected. About 70 per cent of the whole immune system of the human body is located in the bowel – the so-called bowel-associated immune system[93]. Rather, the 'bulk' embodies an interconnected and interdependent ecosystem network, thus validating the term 'satellite'; and, by accounting for function, the term 'environmental crossover' also seems reasonable. To complicate things, we have to remember that events take place without interfering the transit – often considered as the main purpose for the intestine to exist.

In summary, two complex living systems somehow prone to self-organization are to be considered. It is appealing to believe that, by converging, a single tendency may result. At least concerning tangible benefit, this is the case. Finally, there is the possibility that harmony may be counterbalanced by conflict of interest. Fighting for available resources may also affect pathogenic intruders.

POLLUTION: 'MILIEU-SOUP INTERFACE'
A Regulatory Immune link?

The belief in the supposed boundaries provided by the skin may, in fact, be a relic of our earlier simplistic idea that an individual's IDS is unique to that individual. For collective defence, the network has to establish contact with outside milieu. This should have been the case with faulty, crowded dwellings in which inter-body communication is more likely to boost – be it through body fluids or by another means. In fact, direct or indirect immune signalling between individuals may be more common than previously envisaged.

The response to sudden infectious challenges may vary between individuals, but attention should also be paid to reciprocal influence. Previous discussion referred to immune dynamics that, like nerve networks, largely depend on evaluation. For the individual, the acquisition of information about invaders at an early stage may improve the capacity to counteract their virulence. Relevant to us is the fact that the IDS has the means to come into contact with the surrounding environment. Presumably, to extract information out of a *'milieu soup'* requires adaptation.

Almost throughout history, even in times of great success, the bottleneck of packed population centres had been insufficient living space and poor sanitation. Clean water was often not available – folk used to share pans, kettles, spoons, etc. Add pets and livestock, and pollution became critical; no wonder bodies were covered by a *'milieu soup-interface'* film or sheet, permanently supplied by digestive, respiratory and emunctory secretions (a mix of sweat, saliva, mucus, urine, faeces, sperm, etc.). Everything that hosts used to sprinkle, spray or ooze not only has to be seen complemented by commensal flora but also by microbes attracted from outside. A role for molecules, cells or viable fragments from the immune sphere is further to consider.

Now that our way of life undergoes an ever-accelerating path of change, we are far away from prehistory, during which the state of our health tended towards an equilibrium shared by many species. Human creativity was triggered by the advent of history, its unfolding becoming ever accentuated until today, and the outlook suggests that this trend will continue.

Besides considering a *'milieu soup'* as a basis for inter-host immune connection, skin and mucosae has also been briefly discussed as highly selective but also virtual barriers. The increasing diversity of therapeutic principles administered by this route is of note. For the influx of human substance from the neighbourhood, the same may be true. After crossing epithelial/endothelial cell linings in significant amounts, remaining viable could be meaningful. If so, allogeneic gene activation or modification of gene properties in the recipient would not be out of the question.

All in all, this may imply initial conditions with the potential to modulate gene expression – somehow reminiscent of certain pathogenic or even commensal microbial stimuli. As mentioned, the extent to which such external milieu conditions may modulate the meaning of foreign antigens to the IDS is still an open question – they may play a role in individual differences in immune response circuits, signal transduction, etc. Although seemingly implausible in the sparse medium, it may acquire some intrinsic complexity; heterogeneous human/nonhuman shed substance influx will not spare body barriers either. Those molecules, fragments and probably cells that remain viable after the odyssey, could perhaps account for certain effects. The question is whether *à la long* a positive or negative bias in disease transmission could evolve.

Skull Boxes and Skin Bags – Numbers Matter

Could The IDS Display Extracorporeal Activity?

> *"Space comes in degrees of emptiness, but even in the wasteland between galaxies it is not a complete void"*
> Evan Scannapieco *et al.*[94]

Even if the extracorporeal medium is complex, at first sight it does not exert effect on so far known practical aspects from the IDS. But there might indeed be a probability that, in a community in which unhygienic crowded living conditions prevail, much of a supposed immune correlation might the immunity in the medium will relate to the *milieu soup-interface* draped between the bodies almost like a spider's web.

More often than not, a species survival is highly dependent on the community context. Humans evidently use their neural function, centred in the respective skulls, to interconnect. The problem is the other arm: the immune system scattered inside the conventional limits of the body. Think about people talking. Their respective brains do not need to extend their grey mass to establish physical contact. Mainly, the interactive function will be carried out on the basis of waves perceived as light and sound. Technology even allows a conversation with an astronaut on the moon. Distance makes no difference.

We ignore so far how interactive immunity may take place. It is appealing to us that the switch that operates in the case of our neural function may have its immune counterpart, albeit on a much lower scale. For now, molecular and perhaps cellular dynamics may be in charge; it presupposes a dirty congested habitat (see also 'Contrasting Means to Communicate', page 68). In other words, as the tools used by external neural interconnections are totally different from the grey mass in the skull, we can believe that, beyond the body, an equivalent transformation in immunity takes place. On this basis, we would like to be able to explain not just modest observations concerning shanty-town refractoriness to the usual respiratory winter epidemics. It would perhaps explain the survival mystery under besieged conditions. Possibly, phenotype plasticity played a role; it may also be a contributing factor in regulating health in the social context.

An "External Physiology"?

The complex microbial flora settled on the skin varies according local prevalence of secretions, humidity or dryness[95]. A milieu may result that by attracting germs and all kind of microparticles[96,97] may evolve to deserve the term *external physiology*. It is certainly important to explore the spectrum of potential consequences. It is now accepted that abundant interactions with the epithelial surface take place, usually for the better—although nasty consequences cannot be excluded either (for example, impetigo). Clues may provide learning about life in history; to some extent it also may be reflected in today's underprivileged people. In any event, immunity is a main player. Even if molecules only drizzle, there should be some consequence (a metaphor, perhaps, for the Chinese 'water torture'). In the context of complexity, a streamlining tendency to level individuals' immune polymorphism is also a probability. By contrast, it takes time for people not attuned to a particular environment to become familiar with a different inter-human environmental interface and its particular dynamics and dangers (see Figure 9). Finally, when considering the influence that dense *milieus* may exert on individuals, inflammation should not be excluded. Wound healing is an inflammatory process that shares tools with immunity. As such, it deserves careful evaluation under changing inter-host connectivity.

FIGURE 24 ARE THERE MEANS TO READ OTHERS' IMMUNE RESPONSES?

THE 'LOST ENVIRONMENT'

Sweeping body washing is now part of our life. Yet we quickly forget that a large part of humanity still has to live as it did in our unsanitary past. Although shanty-towns often evoke prejudice – we also should have in mind animals mutually attracted by their 'natural' body odour. Only a few generations ago people heavily accounted for body odours' sexual, racial and other connotations. After a successful battle, Napoleon Bonaparte sent a message to his wife, Josephine, ordering her not to take a bath because he would spend the night with her. To link inter-individual immune tuning may not be far-fetched either. The search for an immunological basis for shanty-town kids' refractoriness against common respiratory winter epidemics, increasingly affecting modern society, deserves attention.

107

By improving living space and comfort, current developments prevent *'milieu soup-interfaces'* to endure by replenishment. A putative dual role for the *'lost environment'* has already been discussed. The immune function not only depends on locally interacting molecules and cells; on the other hand, a collective immune responsiveness still is to be demonstrated. It would open the door to a more ample panorama. Brains communicate without getting in physical contact, similarly, immune connections might be disguised by a *'changed face'*.

STARS' ROLE UNCERTAINTY

FIGURE **25** HUMANOIDS, AFTER MILLION-YEAR-SHARIMG THE GLOBE WITH MILLIONS OF SPECIES. DOES OVERWHELMING POWER IMPROVE THE OUTLOOK?

When, under natural conditions, species grow out of control, the contemporaneous equilibrium tends to become destabilized. With regard to humanity, this principle applies not only to the plagues that repeatedly decimated it, but it also applies to the rise of might. *Homo sapiens*, already a star performer, like a tennis champion or a top manager, is becoming increasingly lonely. At his apparent peak, he might soon be made aware that possession of power, without a concomitant care for the abundant species that accompanied him in his long evolution, will rebound on him. Fortunately, we should reiterate that world opinion is heading towards a consensus: already, at the kindergarten, children are taught not to impede Nature's design work.

FIGURE 26 LONELINESS

Obviously, a mere film sheet cannot be seen as competing with senses like sight and hearing, as available to the brain. This abysmal difference, illustrated in Figure 15, means a handicap that obscures the body of functional equivalences. Subjectively at least, nowadays it clouds the function that the *'lost environment'* may have had in historical times.

AMBIGUITY

FIGURE 27 THE OSCILLATION PATTERN

"Yin-Yang expresses the oscillation pattern. Yin-Yang are the two opposed and complementary principles that make up everything. Yin and Yang are the dualistic manifestation of the one and only eternal, indivisible and transcendent principle."

Tao

Leaving aside the nostalgic symbolism expressed by the caption, the scientific establishment is well aware of the subject's significance. In nature, no manifestation can be entirely exempt from ambiguity. Ambiguity never concerns a single entity. It is the relationship between things that, forcibly, implies ambiguity. That relationship is not a static notion; it suggests options – pros and cons. Bearing in mind interdependence, up to a certain degree any behaviour will be shaped by external influences. It is thus not surprising that ambiguity can also be found in the context of superorganism immune co-ordination.

Think about a river that divides into different streams and then becomes one again; similarly, light waves have a potential for such behaviour. In biology, ambiguity is omnipresent, occasionally only as the tip of the iceberg. Immune-status becomes involved as it changes and not infrequently has connotations of ambiguity. It should always be kept in mind when wavering between enhancement and suppression, such as the often surprising progression or regression of autoimmune disease.

By developing the interactive immunity subject within the context of dirt and afar, it is to be expected that a distinct commonness tends to evolve – perhaps closer to equilibrating a chaos-like state. Here again is to remind that we are dealing with uncertainties in the conduct of supposedly sympathetic microbes that belong to the host's flora. Such ambiguity tends to emerge when the host's immune system weakens because of disease, old age or stress. Such opportunities may be exploited by the escorting microbes, which have long proved to be beneficial. As our benign supportive microflora can turn against us, our health problems could worsen. Also, traditional microflora is not exempt from spreading beyond body orifices and skin – this could even affect an entire community.

It might be apposite to say a few words about viruses that have remained harmless or latent for millennia. Even after such a long period, and perhaps because of environmental changes, there is a possibility that the viruses switch from latency or innocuousness to becoming trigger of an epidemic. Such a potential sword of Damocles can mean disaster for billions of people; to keep an eye open for ever-present/latent ambiguity is indeed of paramount importance, and modern man should not consider himself safe from the dangers of plague – nor should he suppose that danger necessarily comes from afar.

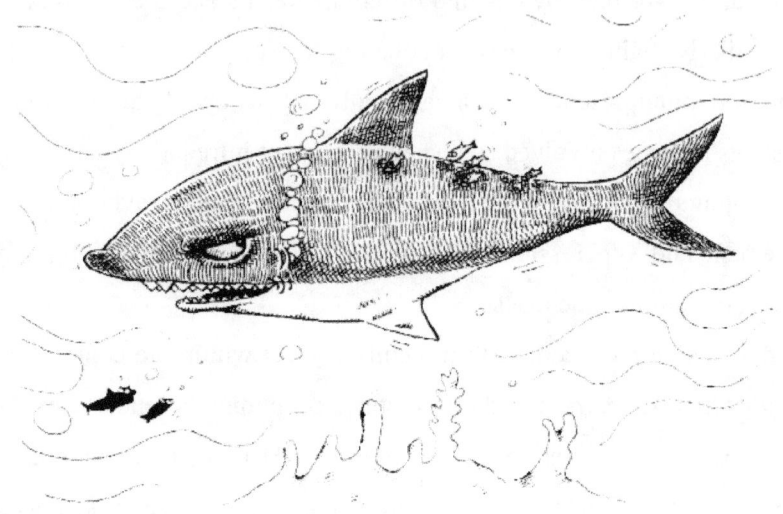

Focusing On Host-Associated Microflora Ambiguity

Immune responsiveness may become activated or inhibited, and some factors may have either effect. In principle, an immune reaction will depend on variables such as initial conditions, milieu, ambiguity (two-way street) and antigen presentation (tridimensional). The subject is complex, going way beyond the scope of the present essay.

Existing theories on the role of the environment do not take into account the diversity of opportunistic bacteria found in a variety of strains within the same species. The emergence of *Escherichia coli* 0157:H7 is a reminder that bacteria may reinvent themselves, adapting to new hosts, new conditions, and new antibiotic countermeasures; these types of bacteria are known as mutators[98]. *E. coli*, some strains of which are essential inhabitants of the intestinal tract, may occur under the form of the highly pathogenic K1 serotype, K1 being nothing but a conglomerate of sialic acid (a sugar molecule), making this germ capable of escaping complement activation[99]. Responsiveness can be modified over time. For example, in HIV infection, independently of the consequences induced by antigenic mutations of the retrovirus, changes in the immune responsiveness of the hosts play an essential role in determining disease progression. Accordingly, the outcome depends on a whole spectrum of antagonistic and synergistic variants.

FIGURE 29 UNITED EFFORTS ARE WELCOME, BUT ASSOCIATION IMPLIES RISKS

The potential for ambiguity in species that belong to the normally associated microflora perhaps best reveals itself in the *Staphylococcus saprophyticus*, a coagulase-negative organism of the normal skin flora. As the name clearly indicates, it was previously considered non-pathogenic. The condition changes when the microbe has occasion to exploit a flaw in the host's defence system. It is now evident that *Staphylococcus saprophyticus* is a main cause of urinary tract infections. Indeed, it is second only to *Escherichia coli* as a cause of cystitis in sexually-active young women. It is also commonly found in elderly men who have undergone urinary tract operations. Systemic bacteraemia is a rare event when upper urinary tract infection occurs.

Since the origins of humans, their lifecycle has been exposed to multiple diseases. Humanity was able to overcome an immense variety of microorganisms in their diverse stages of development and the ways they mutate and evolve – but, in the reciprocal relationship, hosts are often taken by surprise, revealing invaders to be ubiquitous and complex challengers.

One may conclude that ambiguity is inherent in species' ecology and that not only obligate pathogens but also the whole spectrum of innocuousness has the option to become pathogenic.

FIGURE 30 IS IT WISE TO DISMANTLE SUPPORTING FLORA?

CROWDED SHANTY-TOWNS

VERSUS

SPACIOUS HYGIENIC LIVING

PROS AND CONS

♣ Originally 'range of vision', nowadays 'range of comprehension, understanding' – so why not apply to the IDS as well?

III - TRANSFORMATION

"Imperceptibly, even the arch-imperialist was mutating into a little Englander".

Niall Ferguson

HYGIENE: *Gains and Drawbacks*

Hygiene is not a new idea – consider the Old Testament and the habits of the ancient Greeks and Romans – but it was only put into practice in the 19th century after Semmelweiss[♥], simply by hand washing, succeeded in bringing about a drastic reduction in deadly postpartum sepsis.

The currently idealized concept of body hygiene was exported by US troops during the Second World War. For many years Europeans only reluctantly incorporated it into their living habits. During my fellowship at the Freiburg/Br University in 1953, I had opportunity to observe that many people still showed a preference for 'natural' body odours – including in Paris, despite being renowned for its perfumes. Although the supportive role of the usual microbes that populate the surface of the skin often still is questioned, the vulnerability to skin infections observed in America could be related with excessive washing affecting skin flora. For practical purposes, this does not outweigh the benefits of body hygiene, although the disruption of long-standing flora, locally-secreted protective chemicals and even immune-active molecules may also count in contributing to unwanted consequences. Intimate contact with the soil that enriches the body-covering milieu may be a beneficial source of immune maturation. The abundance of Mycobacterium as a non-specific stimulus of cellular immunity, it is to be stressed, belongs to the major incentives for cellular immunity. Good news for many kids that again – almost from birth – are free to roll about in the dirt!

[♥] Ignaz Semmelweiss [1818-1865].

In spite of their handicap, children raised in shanty-towns might harness immune information early – a permanent signal flux could perhaps benefit response tuning and reduce the occurrence of 'immune errors'. Evidence for correlation in community behaviour is emerging in different ways. While most of these creatures meet the criteria of handicapped, they often reveal remarkable resistance to certain seasonal respiratory tract epidemics. Also, these people seem to be exempt from the current trend in autoimmune disease incidence. Rather than with pollution, autoimmune disease is becoming known as a 'disease of hygiene'.

Again, common sense, of course, clearly confirms that the 'other side of the coin' in no way comes close to disqualifying the already historic advance in health benefit that modern hygiene has been able to achieve.

Use it or lose it

In neuroscience, it is not yet clear how learning triggers the survival and even renewal of brain cells in grown-up mammals. Given the parallel with the immune function, applying the idea to a learning process in the context of *community immune interaction*s sounds appealing.

GLOBALIZATION

There is nothing new in the notion that all creatures reveal interdependence patterns, but now the condition is accelerating with no end in sight. Before discussing present anxieties, some anecdotal aspects may illustrate my points.

Wind: organisms, mainly microscopic, but including spiders, mites and insects, have been collected on flights at distances up to 3,000 kilometres from land.

FIGURE 31 AN UNFORGETTABLE INHERITANCE OF THE
HISTORIC ENCOUNTER

Water-flow: many insects depend for food on the organic matter drifting from the upper levels of a river.

Ocean: not only marine invertebrates, but also freshwater animals passively cross oceans by drifting with the currents. The odds are considerably against terrestrial organisms surviving the ordeal, but there are exceptions. In 1892, a remarkable floating island, of an area of 30 square metres, was observed with trees several metres high, having drifted no less than 1,600 kilometres in the Atlantic.

Attachment: animals converted into vectors can disperse organisms over huge areas. For example, mosquitoes have been found to carry a virus many kilometres before infecting a host.

Excrement: contains microbial flora (active metabolism, initially) and may contribute to the dispersal of parasites – even of some plant seeds.

Travel: long-distance flights and shipping of people and objects are the main culprits of globalization.

FIGURE 32 GLOBALIZATION: PRESENT

It must be acknowledged that humanity has been dealing with globalization for a considerable time. The difference now is that contacts are more or less simultaneous across the whole world. The expanded contacts imply demands on immunity.

Since everything is in flux, variables and fluctuations are unclear. Nevertheless, providing that scientific approaches can be perfected, the overall long-term outlook does not appear too bad.

FIGURE **33** PLAGUE

The scene of the immune function is becoming broader. Subliminally, it may perhaps transcend the body and may even have a share in the evolution from incipient connections onwards to more remote networks. An understanding of the intervening period (interregnum) is required.

By the way biologists and ecologists become increasingly familiar with the idea that our planet operates as an integrated network, drawing parallels with the idea of GAIA operating as an organism should perhaps be taken seriously.

FIGURE 34 WHAT SCARES YOU? THE COMPUTER VIRUS OR WORLDWIDE EPIDEMICS AROUND THE CORNER?

Awaiting an Opportunity

Most of the interdisciplinary tools outlined in 'Background' apply to epidemics. Clearly, the subject still causes extreme puzzlement. Sensitive dependence on initial conditions seems an obvious factor, but the early traces in the development of an epidemic disappear. So far, we are waiting for an answer to fundamental questions about the future of a plague. The imprecision built into the calculations rapidly takes over because of the relationship between the particular responses by different hosts. From the point of view displayed put together in the present essay, this may be truth up to a certain point. We have to consider environmental disturbances of diverse nature and, of course, the many pollutants, nowadays largely stemming from consumerism. Factors that alter the normal physiology of pathogens and host-pathogen interactions are numerous. Observations have indicated that a bacterium and its host stimulate each other to carry out successive steps that, in a way, regulate the development of both organisms. People often have to deal with bacteria that apply particularly clever tactics and the question remains as to how to fight them off. It goes without saying that variations in habitat and living conditions do matter. To follow the course of an epidemic, population density and hygiene have to be considered first. Often, differences in the disease's course seem to depend on contradictory signals. Right now is being discussed a finding that nearly two thirds of swine-flu infected people are aged between five and 24, whereas only 1% of cases affected those over 65. It brings immune behaviour with its unforeseen changes into the limelight. Dependence on initial conditions, minimum thresholds, etc., is increasingly being investigated, encompassing a spectrum that runs from firebrand-like progress to complete reversal of the scourge. As already stated, overcrowded living conditions pose questions that should be reassessed. Disease spread, contagion, and recovery pattern are forever changing, thus hampering any informed prognosis. As a result, so far no general agreement for the optimal assessment of the progress of epidemics has been achieved. Here we outline views from two currently diverging authors, Ewald[100] and Wills[101].

Wills sees plagues as temporary blips in an otherwise balanced natural landscape.

'Plagues that afflict us do so because we have upset the balance of nature. The conditions that engender plagues always end sooner or later. This always means that plague pathogens will not survive in the long term unless some of them manage subsequently to lose their virulence and revert to the habits of their low-profile relatives.'

Ewald vehemently opposes these ideas, feeling that Wills, by defining plague as something that eventually subsides, ignores diseases that continue to affect large numbers of people over millennia. Ewald further states that plagues that involve the presence of pathogens in a large number of hosts may taper off in a manner unrelated to any parasite selection for lowered virulence.

'Often plague burns out because enough hosts die so that their population decreases and the pathogen can no longer be transmitted effectively. If the selective pressures of the competing strains favour the more vicious variants, however, the pathogens may maintain high virulence. A low profile may result from a low rate of infection rather than the loss of virulence.'

Some attempts to generalize in epidemiology seem to lead nowhere; so we should not expect a similar long-term evolution for each plague. Such is the case of parasites like the *Plasmodium falciparum* that vectors, anopheles mosquitoes, directly introduce into the bloodstream, thereby providing an entirely passive access to the host. As in the case of malaria no hurdles have to be overcome for penetration, the evolutionary pressure is less. The plasmodium simply has no need of versatility. Very different is the lot of many microbes to which such a direct access is denied. They have to fight and force their way into their prospective host, a process that is often achieved only after fierce competition with resident flora strains. A picture emerges that again reminds us of bullfights. As quotation in the margin, malaria centred in the marshes around the Mediterranean Basin preferentially decimated the normal population, leading to an enrichment of the thalassaemic population.

There are even more schematic points of view; as, for example, the one worked out by Cherkasskii[102], which he presented in ecological terms. By analysing an epidemiological idea that is based on the principles of the systems theory and the information theory, he concludes that the epidemic process is organized by the same principle as living matter with a kind of stability ensured by self-regulation. The author also estimates that human life conditions have been shown to be organically incorporated into the structure of the epidemic process as a regulating subsystem on the socio-ecological level.

The prevalence of epidemics is intimately linked with the adaptive capabilities of species. Here the Darwinian approach must emerge. Transformations that may translate to the epidemics genius are much favoured by ephemeral life-spans of the pathogens[103]. On a lesser scale, the reader is already familiar with the idea that immune defences actually constitute an important part of perplexingly intricate networks that we suppose project to the outer milieu inclusive Altogether, they affect the pros and cons of contagion. The menace inherent in epidemics subjects the immune system to pressures beyond the norm, thereby forcing the individual to seek adaptation or, to put it another way, to learn[104]. It must be suspected that, in the midst of epidemics, people may develop something like immune synchronization, an algebraic sum difficult to elucidate. According to our view, when habitat becomes overpopulated, entanglement of immune networks is probable. Without our knowledge, signalling between individuals may be widespread. We depart from the idea that to become workable, a '*milieu soup-interface*' requires pollution with a critical traffic of appropriate molecules and flora components, possibly in some sort of balance. No doubt much depends on the condition of the hosts and also on the nature of intruders. Should any beneficial effect show up, the immune behaviour of infants might be more receptive to environmental constellations. When maternal protection is fading, until full-blown adult immunity is established, immune bridging and appropriate signalling may be important. We remind the reader to our survey, albeit amateurish, suggesting that, during the usual winter epidemics children from shanty-towns do better at resisting infection than the more privileged children – however, when the shanty-town youngsters do get ill, their handicap as underdogs shows itself as a much worse disease course. An exact knowledge at the molecular level would perhaps provide an interesting perspective.

ASYMMETRY

The IDS can no longer be seen as 'Guardian by Default'.

FIGURE **35** CHANGE IS UNDERWAY

Briefly, millions of known species over the globe display staggering hetero-geneity. Some of them are even found thriving at temperatures near to boiling point, on ice, in ocean depths (over a thousand metres below the surface) or on top of six thousand metre-altitude mountains. The point is that life is a response to what Nature offers: solar energy, atmosphere (oxygen), water, gravity, noise, etc. Charles Darwin, in his treatise *On the Origin of Species by Means of Natural Selection*[105] brought forward guidelines in determining the course of evolution ('Theory of Evolution'), highlighting incessant interspecies and intra-species struggles leading to the prevalence of the most advantageous traits[106]. Fruitful social behaviour, already demonstrated at microbial level, and the building of multi-cellular creatures also make sense. The latter especially rely on sophisticated co-ordination. Specialized tissues and features are supposed to have evolved by long-standing stability, at least inside the body.

1) Muscles that used to be required intensely in many tasks are being replaced by machines. Not too long ago, Napoleon – whilst snow impeded the feeding of his horses – sent the army against Moscow on foot.

2) Nerve stimuli by all-purpose electricity, TV, traffic, etc., prolonged our daily routine.

3) Alimentary industry in tandem with body hygiene substantially reduces the activity of IDS.

4) Present change of living might hinder social immune awareness.

As does any complex system as it is found throughout the natural world, by skipping over the laws of complexity, changes to which human functions and structure are subjected, cannot fully be appreciated. It is to be hoped that the IDS profile will not fall apart, but rather adapt to the forthcoming pace of worldwide change in and around people. Globally, we have survived turbulences that affected our relatively short history, but they may turn pale by comparison, if current transformation will accelerate even more. The impact of events now hatching is hard to predict; already many species have disappeared and to believe the harmony inside our bodies is in trouble does not seem farfetched either. In other words, not only are species at risk, but subtle intra-individual tuning may also be affected by uneven repercussion. Events rather indicate a turning point.

Any asymmetric evolution opens the gates to uncertainty[107]. Obvious and not so obvious estrangements may affect the subtle *status quo* of the immunity/tissue *'détente'*. In fact, most of our view of immune behaviour in the context of organ systems such as musculoskeletal, nervous, digestive tract, etc. had been shaped over a prolonged period, supposedly less unstable. The usual resilience from both the body and community enabled humanity to spread and settle. The question is whether the ongoing impacts and predictable turbulences by imprinting asymmetric tracts will remain inside the frame of the functional window. So far, we have no idea about the forthcoming outcome dynamics – what is clear is that the habitual relationship between body constituents will not be the same as before. Already noteworthy changes took place, but the ever-increasing comfort may also pull the rug; thus, inter-tissue disarray is probable. Common sense tells us that, when species disappear *en masse*, human bodies may not be spared inside. The idea of evolution as a long-term process, even perhaps regulated, may not exactly fit any more either. Evidence for a 'moral bias' to uphold the immune/tissue *'détente'* as known to us, it is to reiterate, also is lacking. So far, no impediment for asymmetric estrangement as promoted by ongoing globalization is in sight. Furthermore, paradoxical observations related to certain micro-organisms and initial malignancy cast doubt about traditional predictability of the immune function[108]. They are to be taken seriously.

While from a clinical point of view parenchyma and immune response mostly combine as expected, divergences have come to light, mainly due to recent insight. As stated, in *sensu stricto*, a specific purpose of one body component to protect others has not been proven; the immune function reveals alien about fate. Ranging from the usual immune-surveillance to tumour escape, the term 'cancer sculpturing' has been coined to define cancer immunoediting in cases in which the immune system favours cancer progression[109].

Asymmetry may also affect autoimmunity. Like any immune response, it is supposed to have been regulated by evolution. Success and failure, the two sides of the coin, rather recently suggests an increasing tendency of autoimmune disease incidence. What has already been coined 'hygiene hypothesis' should perhaps be replaced by a more encompassing term such as, for example, 'contemporaneous asymmetry'. The assumed autoimmune epidemics could be another allusion that harmonic coexistence in the body is already becoming hampered. The IDS may maintain a firm grip on body parts and *vice versa*. Unmasking surrogate views of the immune function may not only help us understand the recent developments of autoimmune disease – in population pockets adapted to 'modern ways of living' especially – it could also perhaps shed light on inter-individual immune relationship – the missing link.

Concerning the tolerance function we referred to innate and adaptive mechanisms doing without harming normal tissues, an 'equilibrium' that still awaits untying, the same as the events that induce a breakdown of immune tolerance. Current changes may perhaps provide an answer. In practical terms, it could be compared to siblings that, when living together, share routines but which obviously, tend to fade the less they have contact in later life.

With reference to the 'portable environments' it seems to us timely to invest in finding out whether the putative immune co-operation supposedly made possible by the '*lost environment*' could be restored without sacrificing the benefits of modern comfort. Here we do not think in terms of individual administered drugs or vaccines, but rather in ways to bridge the means of defence. By succeeding it would perhaps be a complement like iodine added to the water.

Although emphasis centres on estrangement resulting from unpaired changes in the respective phenotype expression, long-standing adaptation will not change so soon and residual effects are to be considered.

CONNECTED   **CLASSIFIED**

FIGURE **36** AN INNER & OUTER IMMUNE FUNCTION?

CONCLUSION

*Paul Ehrlich's 'horror autotoxicus'
stood for almost half a century.*

*Lactic acid, rather than poison, is a valuable source of
muscular energy.*

'Dogmas' expected to be dismissed will not end.

Brains exchange information via electrochemical signals largely converted into waves that we perceive as light or sound. Technology even may allow talk to astronauts, once they land on Mars. But how far the immune system remains isolated in the body we still don't know.

Immune-entangling came into the limelight because of its deterministic chaotic behaviour, which includes modern criteria about unforeseeable changes depending on initial conditions, often at thresholds far below those traditionally thought obtainable. As far as immunity is concerned, a lot of essential elements remain elusive; although, surprisingly, that does not appear to be very obvious in everyday routine in which a relative predictability prevails. We cannot say that there is much difference between immunologists and colleagues dealing with, for example, common wounds. Although extreme situations rarely develop, a 'bombshell' may come at any time, interfering with the natural response sequence: inflammation – fibrosis (variable) – *restitutio in integrum*[110, 111]. What results, intriguingly, is the fact that such a 'bombshell' occurs only exceptionally. It induces us to question whether human society as a whole is being immune-buffered by so far unidentified factors.

Current research on immune repercussions on confinement, be it in support of space flights or stress-conditioned behaviours, mainly concentrates on the mind as a carrier. In such situations, messages reverberate through the

133

recipients' physiological systems. But, in spite of acknowledging the sphere of psychoneuroendoimmunology, to attribute immune interdependence exclusively to an indirect *modus operandi*, any direct immune action being omitted, to say the least, is dubious.

Screening millenary information may be another source of elucidating what for now is believed to have been appalling living conditions. Maybe from historic population centres or from today's shanty-towns key survival factors are brought to light. The know-how that the IDS begins to acquire in the womb – also by breastfeeding – could perhaps be enhanced by unfolding in overcrowded contact. If so, due to the speed in which hygiene and living space is developing, we do not perceive this any longer. By shaking off the body-wrapping sheet, a mix of body secretions and microflora, to which pollutants of all kind are to be added, the consequence will be seen as a *'lost environment'*.

As has been discussed at length, there is no such thing as an insurmountable barrier between subjects. An unhygienic packed milieu ought to provide virtually unlimited antigenic stimuli, often disguised through associations or puzzling camouflage. To be able to thrive, all subjects should be able to distinguish self and non-self. This is part of evolution in which propitious factors gradually promote immunity. But here we postulate that self or non-self targeting does not necessarily derive exclusively from inside conventional body limits – an isolation that seems to oppose biological trends.

Focusing on immunity *per se*, in the body, evolution has created a dispersed network that departs from innate immunity. In humans, the integrated defence system achieved a fine discrimination between a huge number of signals, hostile intruders among them. For the time being we are not aware if this comprehensive system assimilates xenogenic messengers as supportive or detrimental. The question arises whether the whole set of factors – both those already discussed and those yet to be envisaged – could condition some kind of common immune behaviour.

We attempt to make a clear distinction between livings with and without restrictions of space. We know that children change their behaviour when they go to school, as do people when they join the armed forces. Once an individual is locked inside an overcrowded prison, immune performance should also undergo an adaptation process. As often happens in biology, amid disruption stemming from the collapse of normality, compensatory mechanisms may emerge. It may be wrong to generalize, but, in terms of evolution, collective handling appears to be advantageous. Even microbes seek protection, exchanging isolation for integration within a colony. Our immune defences might share a similar process.

Without rejecting the placebo effect outright, we consider the expansion of the immune network, by environment sharing, together with the convergence of flora from the oral, intestinal, genitourinary and skin habitats, plus many organisms that the particular pollution attracted, as factors of potential relevance to the influence of the IDS. As the molecules from one's fellow men become incorporated with one's own, interactions with cells or substances in the host may be of functional significance.

Focusing on microbial environments carried by us, the bulk in the lumen of the colon establishes functional links with the dynamic epithetial layer from the local mucosa, the latter being reinforced by lymphoid tissue rich in phagocytes and immune-molecules-secreting cells. Such encounter is vital to both host and the associated flora. To add are bacteria that managed to descend alive from the oral extremity, either passing through or fighting for a mucosal site.

A minor microbial environment, including about two hundred types of bacteria, settles at the surface of the skin. Interestingly, their distribution differs according to the respective area[95]. Besides substances secreted by the organ skin, the milieu attracts peptides and molecules of many kind — immunoglobulins, receptors, auxiliary molecules and, of course, hormones, neuropeptides among them. It is not exceptional that this 'portable' environment is nourished by industrial pollution as it was by the lack of hygiene in the past. Particles density, by becoming critical may be become a main player that we may call 'milieu soup-interface'. It's significant that the host also releases cells; that is to say that, potentially, via the *'milieu soup'*, the host's IDS network may take bidirectional advantage. We thus face a dynamic film-like sheet that should not be underestimated here.

Finally, at the prospect of a post-antibiotic era, even the faintest indications of a positive counterbalance should be assessed for a so far unaccounted healing effect. To prevent catastrophe with regard to plague, lack of care must be replaced by huge investment. A good start might be to confront the extremes of the Gaussian distribution curve in terms of the conditions in which communities live.

Epidemiological dynamics, expansive or reverting, are of interest. As we have already seen, invaders' dissemination as well as the host's immune behaviour is strongly influenced by environmental changes. Many sources point to the need to redefine the role of the community in patterns of disease transmission and dissemination.

In this context we also referred to the characteristic flare-ups and downs in the course of autoimmune disease and these point to fluctuations in individuals' immune profile with the possibility of similar effects – whether beneficial or adverse– accruing from overcrowding in the host's surroundings.

A whole section of 'Immune Crossover' uses 'trapping' as a common denominator. Autoimmunity is mentioned only marginally here, an exciting subject being microchimerism.

It often happens that important new insights temporarily become overrated and tend to block other proposals, whose importance might not necessarily have been less. For example, if immune interdependence should become a recognized fact, perhaps some role will come to light for those with whom we interact – flora and other bacteria, pets, etc. – perhaps as a complement to the current self-centred exclusivity. It seems essential to precisely locate the individual in the immune environment and *vice versa*.

In summary, the success of immunologists stems from dealing with the internal network context, but the allusions brought forward here suggest that immunity is not necessarily that restricted. Even if available information is fragmentary in the extreme, we envisage that overcrowding in an immune-ecological niche offers the potential for a spectrum of consequences for good or for bad. Adaptive immunological implications for overcrowding have yet to be credited.

Looking for new insights, also inspired by interdisciplinary thoughts, to integrate modern ideas such as complexity, self-organization and co-evolution may help to realize the meaning of 'our portable' environments.

Perhaps we will now see a change in the human predatory habits, although it could well take a considerable amount of time. It is, however, exciting to witness the enthusiasm with which young people assume responsibility, very far from the classical assumption that they are only either innocent or cruel.

EPILOGUE

While we have the feeling that our 'thought experiment' could bear fruits, we are aware that there is not such a thing as an axis *'brain – immune system'*. Although main players, they also depend on the regulation by the rest of the body to which, in a social context, even more factors need to be accounted for. Complexity being all around; deep feelings of love are not exempt from all kinds of inward and outward interferences. As Erich Kästner[1] puts it citing a French poet's gloomy allegory about the hopelessness of the lovers' perfect encounter[2]; they in fact remain isolated, immobilized and blindfolded in a crude burlap bag. In other words, networks transcendence notwithstanding, full amalgamation is utopia.

---

[1] German poet and satirist—children's literature (1899 – 1974).
[2] no name given

A SUCCINCT GLOSSARY

"Adaptation": one of the basic phenomena of biology. It is the process whereby an organism becomes better suited to its habitat. Also, the term *adaptation* may refer to a characteristic which is especially important for an organism's survival. Such adaptations are produced in a variable population by the better suited forms reproducing more successfully, that is, by natural selection.

"Adjuvant": a substance which enhances a response.

"Antibody": a serum protein adapted to specific binding to the immunizing epitope.

"Antigen": a material with potential to induce a specific immune response.

"Antigen-presenting cell" (APC): a cell that expresses on its membrane molecules involved in processing and presentation of antigen.

"Apoptosis": regulated cell death that usually does not affect counterparts.

"Arborvitae": used as a decorative motif, probably of ultimately Assyrian origin.

"Atopy": a genetically determined state of hypersensitivity to environmental allergens Type I allergic reaction is associated with the IgE antibody and a group of diseases such as hay fever, asthma and atopic dermatitis.

"Attunement": sees the human body as a dynamic holistic self-healing expression of a deeper spiritual self.

"B cell": a lymphocyte cell type destined to produce, to carry and to release immunoglobulin; it also expresses class II MHC on its surface.

"cAMP": cyclic adenosine monophosphate, a cyclic nucleotide, binds to protein kinases inducing phosphorylation, thereby activating various enzymes. Intracellular mediator of the action of distinct hormones at cell membrane.

"Catalysis": increase in the velocity of a chemical reaction or process that is consumed by the presence of a substance that is not consumed in the net chemical process.

"Chimera": refers to genetically distinct cells of an individual. The term derives from the myth of an animal possessing the head of a lion, the body of a goat and the tail of a snake.

"Chronon": a nondecomposable time interval unit related to a particular issue.

"Class I, II, and III molecules": proteins encoded by genes in the MHC.

"Complement": a number of serum proteins that activate in a cascade order that may be triggered either 'classically' by the interaction of an antibody with a specific antigen or an alternative (shorter) pathway through nonspecific stimuli.

"Co-stimulation": a co-stimulatory molecule is often crucial to the development of an effective immune response.

"Cyclic (periodic) phenomena": in living organisms, also biological rhythms ('*chronobiology*' chrono = time).

"Cytokines": cell-secreted messenger molecules acting in cells' 'talk'.

"Dendritic cells (DC)": Immune cells that form part of the mammalian immune system. They are present in small quantities in tissues that are in contact with the external environment, mainly the skin and the inner lining of the nose, lungs, stomach and intestines.

"Direct antiglobulin test (DAT)": to detect antibodies fixed on the red cell membrane.

"Entropy": the measure of that part of the heat or energy of a system which is not available to perform any work.

"Epitope": an antigenic determinant – ligates to a specific antibody.

"Eustress": positive stress; physiological optimum inherent to active living without being exposed to particular aggression. Eustress promotes restoring health.

"Gaia": comes from the Greek Goddess of the Earth who gave birth to the sky, mountains and sea. It implies the fragile interaction and interdependence that links all things – especially those related to biology. It is ever more convincing that autoregulatory homeostasis by the earth is to be seen as a contributory factor to the long-term stability of Earth in terms of our solar system.

"Gaussian distribution curve": bell-shaped symmetrical frequency distribution for a set of variable data.

"Habituation": learning to disregard stimuli that are without significance to a living being. It may be regarded as a fundamental property in biology.

"Haplotype": a set of alleles of a group of closely linked genes, such as the HLA complex which is usually inherited as a unit.

"Heat-shock proteins" (HSP): the function of the group of proteins is virtually similar in all living organisms, from bacteria to humans. 'Cellular stress response', also called 'heat-shock response', appears when the cell is under stress, such as heat or others. HSP also occur under non-stressful conditions. Belonging to a cell's repair system, they may dismiss old proteins and support the folding of newly synthesized ones.

"HIV": human immunodeficiency virus.

"HLA" (human leukocyte antigens): histocompatibility antigens governed by the human major histocompatibility complex (MHC).

"Idiotype": the antigenic determinant (idiotope) in the variable region of an antibody.

"Idiotypic-antiidiotypic network": to bind an antigenic determinant (epitope), an immunoglobulin molecule needs to adapt by somatic mutation; thereby it may become an antigenic target. Antiidiotype is the corresponding antibody; progression is network alike.

"Integrated defence system" (IDS): rather than the 'immune system' (IS), this presupposes versatility; besides hormones and nerves, *in sensu stricto*, most body components are involved.

"Integrin": a member of a family of cell surface adhesion molecules that bind mainly to extracellular matrix.

"Interferon(s)" (IFN): a group of body-secreted proteins having antiviral and immunomodulatory properties, e.g. IFN-gamma (IFNγ) tends to proinflammatory effects.

"Interleukins": glycoproteins produced by diverse white blood cells that mainly react with other leukocytes – promoting or attenuating inflammatory reaction (from IL1 to IL20 so far).

"Isotype": Any of the subclasses of immunoglobulins defined by the chemical and antigenic characteristics of their constant regions.

"Limbic system": term loosely applied to a group of brain structures involved with autonomic (independent) functions, certain aspects of emotion and behaviour and with other activities.

"Lypopolysaccharides" (LPS): a component of Gram-negative bacteria that induce 'trigger-happy' host reactions.

"Meme": 'an idea, behaviour, or style usage that spreads from person to person or within a culture' (Webster's Dictionary).

"Major histocompatibility complex class I" molecules (MHC I): take part in antigen presentation to cytotoxic T (CD8+) cells.

"Major histocompatibility complex class II" molecules (MHC II): take part in antigen presentation – to helper T (CD4+) cells.

"Mutation": is a change in the chemistry of a gene that is perpetuated in any subsequent divisions of the cell in which it occurs.

"Naïve lymphocytes": lymphocytes that have developed but have yet to be activated by antigen.

"Noninteger dimensionality": differs from relations between natural numbers (1, 2, 3, etc.).

"PCR": the polymerase chain reaction is a basic tool in molecular genetics; DNA PCR is most efficient when amplifying selected sections of DNA to produce multiple molecular copies that allow the analysis of any short sequence of DNA (or RNA) without the need for weeks of cloning in bacteria.

"Phenotype": the observable physical or biochemical characteristics of an organism, as determined by both genetic make-up and environmental influences. It differs from **"genotype"**, which means only the genetic make-up (adapted from Wikipedia).

"Pheromones": pherein [Greek *pheros*, far], transporter & hormone, exciter [*horman*, to excite]: chemicals emitted by one individual to evoke some behaviour in another of the same species, and also for self-orientation.

"Placebo effect": the notion at least dates as far back as Hippocrates, who observed that certain gravely ill people seemed to recover through sheer 'contentment' with their doctors.

"Probiotics": commensal bacterial species with beneficial physiologic or therapeutic activities.

"Psychoneuroendoimmunology" ("Neuroendoimmunology"): the holistic approach of interdependent mechanisms.

"Selectins": cell-surface adhesion molecules diversely binding carbohydrates.

"Somatic mutation": is a mutation occurring in the general body cells – rather than to the germ cells – and is not transmitted to progeny.

"Superantigen": an immunostimulatory molecule produced by bacteria and viruses. The characteristics of binding to the MHC outside the antigen-binding cleft, and of stimulating T cells in a T-cell receptor, provide potent immune effects.

"T cell": lymphocyte cell types that are mostly removed or trained to differentiate in the thymus. Specific maturation requires an encounter with peptides presented mostly by dendritic cells. 'Helper T cells' (Th) help trigger B cells, generate cytotoxic T cells, etc.

"Teleonomy": doctrine that attributes a sense to the different biological properties.

"Tolerogens": antigens with the capacity to induce tolerance.

"Tumour necrosis factor(s)" (TNF): proinflammatory cytokine.

REFERENCES

1. Nee, S. Extinction, slime, and bottoms. *PLoS Biol.* 2004; 2: E272.

2. Garrett, L. *The Coming Plague.* New York: Penguin Books, 1994.

3. Grady, D. Quick-change pathogens gain an evolutionary edge. *Science.* 1996; 274: 1081.

4. Ames, B. N. , Profet, M. & Gold, L. S. Dietary pesticides (99.99% all natural). *Proc. Natl Acad. Sci. USA.* 1990; 87: 7777–7781.

5. Boroviczény, C. G. (ed). Standardization in Haematology. *Bibl. Haemat.* 1966; 24: 186–193.

6. Nowak, M. A. & Bangham, C. R. Population dynamics of immune responses to persistent viruses. *Science.* 1996; 272: 74-79.

7. Kapra, F. *The Web of Life.* Harper Collins Publishers. 1996.

8. Devaney, R. *An Introduction to Chaotic Dynamical Systems.* Westview Press. 2003.

9. Gleick, J. Chaos: *Making a New Science.* Penguin. 1988.

10. Wolfe, N. (from *New York Times* article: Deep in the Rain Forest, Stalking the Next Pandemic. 2008)

11. Turney, C. *Bones, Rocks and Stars: The Science of When Things Happened.* Palgrave Macmillan. 2007.

12. Turney, C. *Ice, Mud and Blood: Lessons from Climates Past.* Palgrave Macmillan. 2008.

13. Heijnen, C. J. Receptor regulation in neuroendocrine-immune communication: current knowledge and future perspectives. *Brain Behav. Immun.* 2007; 21: 1–8.

14. Stockhorst, U. & Klosterhalfen, S. Conditioning mechanisms and psychoneuro-immunology (in German). *Psychother. Psychosom. Med. Psychol.* 2005; 55: 5–19.

15. Hangalapura, B. N., Nieuwland, M. G., de Vries Reilingh, G., van den Brand, H., Kemp, B. & Parmentier, H. K. Durations of cold stress modulates overall immunity of chicken lines divergently selected for antibody responses. *Poult. Sci.* 2004; 83: 765–775.

16. Cohen, S, Tyrrell. D. A. & Smith, A. P. Psychological stress and susceptibility to the common cold. *N. Engl. J. Med.* 1991; 325: 606–612.

17. Melchior, M., Niedhammer, I., Berkman, L. F. & Goldberg, M. Do psychosocial work factors and social relations exert independent effects on sickness absence? A six year prospective study of the GAZEL cohort. *J. Epidemiol. Community Health.* 2003; 57: 285–293.

18. Hennessy, M. B. Social influences on endocrine activity in guinea pigs, with comparisons to findings in nonhuman primates. *Neurosci. Biobehav. Rev.* 1999; 23: 687–698.

19. Gust, D. A., Gordon, T. P., Brodie, A. R. & McClure, H. M. Effect of companions in modulating stress associated with new group formation in juvenile rhesus macaques. *Physiol. Behav.* 1996; 59: 941–945.

20. Kingston, S. G. & Hoffman-Goetz, L. Effect of environmental enrichment and housing density on immune system reactivity to acute exercise stress. *Physiol. Behav.* 1996; 60: 145–150.

21. Bennett, M. P. & Lengacher, C. Humor and Laughter May Influence Health. Evid Based Complement. *Alternat. Med.* 3(2): 187–190. 2006.

22. Flavius Josephus. *The Jewish War*. Harvard University Press.

23 Harpending, H., Hawks, J. *Scientific American* Jan. 2009.

24. Margulis, L. *Symbiosis in Cell Evolution*. W. H. Freeman & Company. 1981.

25 Cairns, J. *et.al.* The origin of mutants. *Nature.* 1988; 335: 142–145.

26. Poo, J. I., Martin, P. I., Rewald, E. Autoimmunity and time entanglement. *Ann. N. Y. Acad. Sci.* 2007; 1109: 37–39.

27. Martín, P. I., Malizia, A. I., Rewald, E. A propos time and autoimmunity. *Clin. Rev. Allergy Immunol.* 2008; 34: 380–394.

28. Garland, T. Jr., Kelly, S. A. Phenotypic plasticity and experimental evolution. *J. Exp. Biol.* 2006 Jun; 209 (Pt 12): 2344–2361.

29. Kitano, H. Self-Extending Symbiosis: A Mechanism for Increasing Robustness Through Evolution. *Biological Theory.* 2006; 1: 61–66.

30. Aharoni, A., Gaidukov, L., Khersonsky, O,, Gould, McQ. S., Roodveldt, C. & Tawfik, S. The evolvability of promiscuous protein functions. *Nature Genetics* 2005; 37: 73–76.

31. Lovelock, J. Gaia: the living Earth. *Nature.* 2003; 426: 769–770.

32 Aubrey, J. *A Brief Life of Thomas Hobbes*. Nonpareil Books, No 77.

33. Mckeon, R. *The Philosophy of Spinoza: the Unity of his Thought.* Woodbridge, Conn.: Ox Bow Press, 1987.

34. Gaardner, J. *Sophie's World*. London: Phoenix House. 1995.

35. Levin, D. M. & Solomon, G. F. The discursive formation of the body in the history of medicine. *J. Med. Philos.* 1990; 15: 515–537.

36. Ventegodt, S., Flensborg-Madsen, T., Andersen, N. J. & Merrick, J. The life mission theory VII. Theory of existential (Antonovsky) coherence: a theory of quality of life, health, and ability for use in holistic medicine. *Sci. World J.* 2005;5: 377–389.

37. Traniello, J. F., Rosengaus, R. B., Savoie, K. The development of immunity in a social insect: evidence for the group facilitation of disease resistance. *Proc. Natl. Acad. Sci. U. S. A.* 2002; 99 (10): 6838–6842.

38. Levin, A. S. & Byer, V. S. Environmental illness: a disorder of immune regulation. *Occup. Med.* 1987; 2: 669–681.

39. Weiss, A. & Littman, D. R. Signal transduction by lymphocyte antigen receptors. *Cell.* 1994; 76: 263–274.

146

40. Kaul, R., Rutherford, J., Rowland-Jones, S. L., Kimani, J., Onyango, J. I., Fowke, K., MacDonald, K., Bwayo, J. J., McMichael, A. J. & Plummer, F. A. HIV-1 Env-specific cytotoxic T-lymphocyte responses in exposed, uninfected Kenyan sex workers: a prospective analysis. *AIDS* 2004; 18: 2087–2089.

41. Bäckhed, F., Ley, R. E., Sonnenburg, J. L., Peterson, D. A. & Gordon, J. I. Host-Bacterial Mutualism in the Human Intestine. *Science*. 2005; 307: 1915–1920.

42. Nathan, C. Points of control in inflammation. *Nature*. 2002: 26; 420: 846–852.

43. Kaneko, Y., Nimmerjahn, F. & Ravetch, J. V. Anti-inflammatory activity of immunoglobulin G resulting from Fc sialylation. *Science*. 2006; 313: 670–673.

44. Adcok, F. E. *Thucydides and his History*. Cambridge: Cambridge University Press. 1963.

45. Coutinho, A. Will the idiotypic network help to solve natural tolerance? *Trends Immunol*. 2003; 24: 53–54.

46. Lorenzo, M., Sánchez, P. & Rewald, E. Noise and microflora behavior. *Ann. NY Acad. Sci*. 2004; 1029: 358–360.

47. Sanchez, P., Lorenzo, M. & Rewald, E. Stochastic resonance-like phenomena in the context of the alimentary tract. *Ann. NY Acad. Sci*. Dec. 2004; 1029: 390–393.

48. Smyth, L. A., Afzali, B., Tsang, J., Lombardi, G., Lechler, R. I. Intercellular transfer of MHC and immunological molecules: molecular mechanisms and biological significance. *Am. J. Transplant*. 2007 Jun; 7 (6): 1442–1449.

49. Benjamín, E. & Leskowitz, S. *Immunology*, 2nd edition. New York: Wiley-Liss, 1991: 302–309.

50. Davies, H. *Introductory Immunology*. London: Chapman & Hall. 1997; 258–270.

51. Billingham, R. E., Brent, L., Medawar, P. B. Actively acquired tolerance of foreign cells. *Nature* 1953; 172: 603–606.

52. Jiga, L. P., Bauer, T. M., Chuang, J. J., Opelz, G., Terness, P. Generation of tolerogenic dendritic cells by treatment with mitomycin C: inhibition of allogeneic T-cell response is mediated by downregulation of ICAM-1, CD80, and CD86. *Transplantation* 2004; 77: 1761–1764.

53. Jiga, L. P., Ehser, S., Kleist, C., Opelz, G., Terness, P. Inhibition of heart allograft rejection with mitomycin C-treated donor dendritic cells. *Transplantation* 2007; 83: 347–350.

54. Terness, P., Kleist, C., Simon, H., Sandra-Petrescu, F., Ehser, S., Chuang, J. J., Mohr, E., Jiga, L., Greil, J., Opelz, G. Mitomycin C-treated antigen-presenting cells as a tool for control of allograft rejection and autoimmunity: From bench to bedside. *Human Immunology* (in press)(electronic internet pre-print).

55. Steinman, R. M., Hawiger, D., Nussenzweig, M. C. Tolerogenic dendritic cells. *Annu. Rev. Immunol*. 2003; 21: 685–711.

56. Nimchuk, Z., Eulgem, T., Holt 3rd, B. F. & Dangl, J. L. Recognition and response in the plant immune system. *Annu. Rev. Genet*. 2003; 37: 579–609.

57. Rewald, E., Francischetti, M. M. Defense system shortcuts and limits of scope. *Med. Hypotheses*. 2000; 55: 277–282.

58. Pawlik, J. R. Marine invertebrate chemical defences. *Chem. Rev.* 1993; 93: 1911–1922.

59. Pophof, B. Moth pheromone binding proteins contribute to the excitation of olfactory receptor cells. *Naturwissenschaften.* 2002; 11: 515–518.

60. Justus, K. A., Carde, R. T. & French, A. S. Dynamic properties of antennal responses to pheromone in two moth species. *J. Neurophysiol.* 2005; 93: 2233–2239.

61. Bardwell, L. A walk-through of the yeast mating pheromone response pathway. *Peptides.* 2005: 339–350.

62. Baird, R. C., Johari, H. & Jumper, G. Y. Numerical simulation of environment modulation of chemical signal structure and odor dispersal in the open ocean. *Chem. Senses.* 1996; 21: 121–134.

63. Kapoor, B. G. & Khanna, B. *Ichthyology Handbook.* Springer. 2004.

64. Koch, A. L. Genetic response of microbes to extreme challenges. *J. Theor. Biol.* 1993; 160: 1–21.

65. Waters, C. M. & Bassler, B. L. Quorum Sensing: Cell-to-Cell Communication in Bacteria. *Annu. Rev. Cell Dev. Biol.* 2005; 21: 319–346.

66. Dong, Y. H. & Zhang, L. H. Quorum sensing and quorum-quenching enzymes. *J. Microbiol.* 2005; 43 Spec No: 101–109.

67. Hong, S., Belanger, M., Whitlock, J., Kozarov, E. & Progulske-Fox, A. Hemagglutinin B is involved in the adherence of *Porphyromonas gingivalis* to human coronary artery endothelial cells. *Infect. Immune.* 2005, 73: 7267–7273.

68. Pallasch, T. J. Antibiotic resistance. *Dent. Clin. North Am.* 2003; 47: 623–639.

69. Nydegger, U. E. In video: Chasing *After the Molecule.* Rewald, E. 1995, Vienna: Good Day Film Production, Kräftner & Wagenhofer.

70. Artlett, C. M. Pathophysiology of fetal microchimeric cells. *Clin. Chim. Acta.* 2005; 360: 1–8.

71. Sarkar, K. & Miller, F. W. Possible roles and determinants of microchimerism in autoimmune and other disorders. *Autoimmun. Rev.* 2004; 3: 454–463.

72. Chen, R., Xiong, S., Yang, Y., Fu, W., Wang, Y. & Ge, J. The relationship between human cytomegalovirus infection and atherosclerosis development. *Mol. Cell Biochem.* 2003; 249: 91–96.

73. Mosorin, M., Surcel, H. M., Laurila, A., Lehtinen, M., Karttunen, R., Juvonen, J., Paavonen, J., Morrison, R. P., Saikku, P. & Juvonen, T. Detection of Chlamydia pneumoniae-reactive T lymphocytes in human atherosclerotic plaques of carotid artery. *Arterioscler. Thromb. Vasc. Biol.* 2000; 20: 1061–1067.

74. Theofilopoulos, A. N. The basis of autoimmunity: Part II. Genetic predisposition. *Immunology Today.* 1995; 16: 150–159.

75. Braun-Fahrlander, C., Riedler, J., Herz, U., Eder, W., Waser, M., Grize, L., Maisch, S., Carr, D., Gerlach, F., Bufe, A., Lauener, R. P., Schierl, R., Renz, H., Nowak, D.& von Mutius, E; Allergy and Endotoxin Study Team. Environmental exposure to endotoxin and its relation to asthma in school-age children. *N. Engl. J. Med.* 2002; 347: 869–877.

148

76. Weiss, R. A., McMichael, A. J. Social and environmental risk factors in the emergence of infectious diseases. *Nat. Med.* 2004; 10: S70–S76.

77. Bach, J. F. The effect of infections on susceptibility to autoimmune and allergic diseases. *N. Engl. J. Med.* 2002; 347: 911–920.

78. Chen Jr, P. C., Chen, D., Chen, P. D., Tuen, M., Cohen, S., Migueles, S. A., Connors, M., Rosenberg, E., Malhotra, U., Gonzalez, C. & Hioe, C. E. HIV-1-infected patients with envelope-specific lymphoproliferation or long-term nonprogression lack antibodies suppressing glycoprotein 120 antigen presentation. *J. Infect Dis.* 2004; 189: 852–861.

79. Connors, M. M. The politics of marginalization: the appropriation of AIDS prevention messages among injection drug users. *Cult. Med. Psychiatry.* 1995; 19: 425–452.

80. Massabki, P. S., Accetturi, C., Nishie, I. A., da Silva, N. P., Sato, E. I. & Andrade, L. E. C. Clinical implications of autoantibodies in HIV infection. *AIDS.* 1997; 11: 1845–1850.

81. González, C. A. Semiología Inmunohematológica en la Infección por el VIH-1. *Rev. Argent. Infectología.* 1997; 10: 11–20.

82. González, C. A., Guzmán, L. & Nocetti, G. Drug-dependent antibodies with immune hemolytic anemia in AIDS patients. *Immunohematology.* 2003; 19: 10–15.

83. González, C. A. Successful treatment of autoimmune hemolytic anemia with intravenous immunoglobulin in a patient with AIDS. *Transplant Proc.* 1998; 30: 4151–4152.

84. Centers for Disease Control (CDC). Human immunodeficiency virus (HIV) infection codes. *MMWR.* 1987; 36S: 1–20S.

85. Lai, M., d'Onofrio, G., Visconti, E., Tamburrini, E., Cauda, R., Leone, G. Aetiological factors related to a positive direct antiglobulin test result in human immunodeficiency virus-infected patients. *Vox Sang.* 2006 May; 90 (4): 325–330.

86. Hofstadter, D. Cited by *The Guardian* (August 8), London. 1997.

87. Rewald, E., Francischetti, M. M. & González, C. No man is an island; no island is an island: does the immune network extend beyond the limits of skin? *Med. Hypotheses.* 1999; 52: 325–327.

88. André, J.-B., Gupta, S., Frank, S. & Tibayrenc, M. Evolution and immunology of infectious diseases: what's new? *Infection, Genetics and Evolution.* 2004; 4: 69–75.

89. Cohen, I. R. The cognitive paradigm and the immunological homunculus. *Immunol. Today.* 1992; 13: 490–494.

90. Stappenbeck, T., Hooper, L. & Gordon, J. Developmental regulation of intestinal angiogenesis by indigenous microbes via Paneth cells. *PNAS.* 2002; 24: 15451–15455.

91. Weiner, H. L., Mayer, L. & Strober, W. (eds.). Oral Tolerance – New Insights for Clinical Application. *Ann. New York Acad. Sci.* 1994; vol. 1029.

92. Macdonald, T. T. & Monteleone, G. Immunity, inflammation, and allergy in the gut. *Science.* 2005; 307: 1920–1925.

93. When the biofilm is broken. Inflammation at Interfaces. 2006. Online Pub.

94. Scannapieco, E., Petitjean, P. & Broadhurst, T. The emptiest places. *Sci. Am.* 2002; 287: 56–63.

95. Grice, E. A. *et al.* Topographical and Temporal Diversity of the Human Skin Microbiome. *Science* May 2009; 324 (5931): 1190–1192.

96. Rewald, E., Malizia, A. I., Martín, P. I. The other side of the coin: could microparticles serve as an interindividual immune link? *Scand. J. Immunol.* 2008; 67: 103. Comment on: *Scand. J. Immunol.* 2007; 66: 159-65.

97. Martín, PI., Sánchez, P. A., Rewald, E. Microparticles and the Hygiene Hypothesis. *Ann. NY Acad. Sci.* 2009; 1173:409-421.

98. Tlaskalova-Hogenova, H., Stepankova, R., Hudcovic, T., Tuckova, L., Cukrowska, B., Lodinova-Zadnikova, R., Kozakova, H., Rossmann, P., Bartova, J., Sokol, D., Funda, D. P., Borovska, D., Rehakova, Z., Sinkora, J., Hofman, J., Drastich, P. & Kokesova, A. Commensal bacteria (normal microflora), mucosal immunity and chronic inflammatory and autoimmune diseases. *Immunol. Lett.* 2004; 93: 97–108.

99. Sansonetti, P. J. War and peace at mucosal surfaces. *Nature Reviews Immunology.* 2004; 4: 953–964.

100. Ewald, P. *Evolution of Infectious Disease.* Oxford: Oxford University Press. 1994.

101. Wills, C. *Yellow Fever – Black Goddess: the Coevolution of Plague.* London, Addison-Wesley. 1996.

102. Cherkasskii, B. L. The epidemic process as a system. I. The structure of the epidemic process (in Russian). *Zh. Mikrobiol. Epidemiol. Immunobiol.* 1985; 3: 45–51.

103. Coutinho, A. The Le Douarin phenomenon: a shift in the paradigm of developmental self-tolerance. *Int. J. Dev. Biol.* 2005; 49: 131–136.

104. Le Souëf, P. N., Goldblatt, J. & Lynch, N. R. Evolutionary adaptation of inflammatory immune responses in human beings. *Lancet.* 2000; 356: 242–244.

105. Darwin, C. *On the Origin of Species by Means of Natural Selection.* 1859.

106. Phillips, R. E. Immunology taught by Darwin. *Nat. Immunol.* 2002; 3: 987–989.

107. Rewald. E., Barberis, L. & Francischetti, M. M. Asymmetry alters immune-partner relationship. *Cell. Mol. Biol.* (Noisy-le-grand). 2004; 50 Online Pub: OL515–516.

108. Orange, J. S., Fassett, M. S., Koopman, L. A., Boyson, J. E. & Strominger, J. L. Viral evasion of natural killer cells. *Nat. Immunol.* 2002; 3: 1006–1012.

109. Dunn, G. P., Old, L.J. & Schreiber, R. D. The three Es of cancer immunoediting. *Annu. Rev. Immunol.* 2004; 22: 329–360.

110. Rewald, E. IVIG as a nonspecific accelerator of restitutio ad integrum. *Transfus. Sci.* 2000; 23: 163–164.

111. Rewald, E., Francischetti, M. M. & Nydegger, U. E. IVIG-pools: regulatory gifts – transiting from harmony toward harmonious immunoglobulins: why? and why not? *Transfus. Apher. Sci.* 2001; 25: 113–137.

112. Benaroya-Milshtein N., Apter A., Yaniv I., Kukulansky T., Raz N., Haberman Y., Halpert H., Pick C. G., Hollander N. Environmental Enrichment Augments the Efficacy of Idiotype Vaccination for B-cell Lymphoma. J Immunother 2007; 30: 517–522

INDEX

www.ingramcontent.com/pod-product-compliance
Lightning Source LLC
Chambersburg PA
CBHW060850170526
45158CB00001B/295